Strategic Engineering for Cloud Computing and Big Data Analytics

Amin Hosseinian-Far · Muthu Ramachandran
Dilshad Sarwar
Editors

Strategic Engineering for Cloud Computing and Big Data Analytics

 Springer

Editors
Amin Hosseinian-Far
School of Computing, Creative
  Technologies and Engineering
Leeds Beckett University
Leeds
UK

Dilshad Sarwar
School of Computing, Creative
  Technologies and Engineering
Leeds Beckett University
Leeds
UK

Muthu Ramachandran
School of Computing, Creative
  Technologies and Engineering
Leeds Beckett University
Leeds
UK

ISBN 978-3-319-84915-7   ISBN 978-3-319-52491-7   (eBook)
DOI 10.1007/978-3-319-52491-7

This Springer imprint is published by Springer Nature
The registered company is Springer International Publishing AG
The registered company address is: Gewerbestrasse 11, 6330 Cham, Switzerland

Foreword

The first time I got involved in a rigorous problem-solving effort related to system resilience and strategic engineering was during my internship at Intel, Santa Clara. As a Ph.D. student, my main concern at school was to build functional chips, chips that simply work as I powered them up. But at Intel Skylake Server CPU group, for the first time, I was facing questions about the reliability of the server chip, the back-end platform of cloud computing and big data management, in an uncertain, perhaps distant future. The following questions were the main considerations:

- How can we foresee the failures and avoid (or delay) them during the design stage?
- What type of failure is more likely to happen in a specific block of the system?
- What are Mean Time to Failure (MTTF) and Mean Time Between Failures (MTBF)?
- How can we maintain and dynamically correct our system while it is running?
- How can we expand and scale the system with new software and hardware features without jeopardising reliability, sustainability and security?

The short exposure to strategic engineering had a long lasting impact on my approach toward engineering in general and integrated circuits and systems design in particular. Later that autumn, when I returned to my tiny damp cubicle at building 38 of MIT to continue working on nano-relay based digital circuits, my concern was no longer merely the functionality of my systems right out of the box. The durability, scalability, resilience and sustainability of the system started to play an important role in my design strategies and decisions.

In the new age of global interconnectivity, big data and cloud computing, this book provides a great introduction to the flourishing research field of strategic engineering for cloud computing and big data analytics. It encompasses quite a few interesting topics in this multidisciplinary research area and tries to address critical questions about systems lifecycle, maintenance strategies for deteriorating systems, integrated design with multiple interacting subsystems, systems modelling and analysis for cloud computing, software reliability and maintenance, cloud security and strategic approach to cloud computing.

While many questions about the future of big data in the next 20 years are unanswered today, a good insight into the computational system modelling, maintenance strategies, fault tolerance, dynamic evaluation and correction and cloud security would definitely pave the way for a better understanding of the complexity of the field and an educated prediction of its future.

Dr. Hossein Fariborzi
Assistant Professor
King Abdullah University of Science
and Technology
Saudi Arabia

Preface

This first edition of Strategic Engineering for Cloud Computing and Big Data Analytics focuses on addressing numerous and complex, inter-related issues which are inherently linked to systems engineering, cloud computing and big data analytics. Individuals have consistently strived through engineering and technology to improve the environment on a global scale. With this ever-changing societal environment, there are far greater challenges which are required to address these phenomenal technological evolutionary demands.

The primary audience for the book is research students, industry experts and researchers in both industry and academia, masters level students, undergraduate students who are interested in the subject area with a view of gaining greater understanding and insight in the strategic implications of cloud computing in terms of big data analytics additionally managers wishing to gain a better understanding of introducing and implementing new improved technology concepts within their organisations. This book is particularly relevant for readers wishing to gain an insight into the overall constructs of systems engineering in line with the growing dimensions of cloud and big data analytics. It covers a wide range of theories, techniques, concepts, frameworks and applied case studies related to key strategic systems development, maintenance and modelling techniques.

The subject of strategic engineering is far too complex for such simple solutions and therefore the book provides a critical and reflective systems thinking approach. The book is particularly useful in illustrating an opulent foundation of materials which clearly and objectively draw upon a number of examples and real-world case studies in order to demonstrate the many key issues facing the ever-changing technological environment we live in today.

There are three key parts the book focuses on. Part I focuses on 'Systems Lifecycle, Sustainability, Complexity, Safety and Security'; Part II focuses on 'Systemic Modelling, Analysis and Design for Cloud Computing and Big Data Analytics' and the final Part III focuses on 'Cloud Services, Big Data Analytics and Business Process Modelling', focusing on strategic approaches, with the onus on cloud services and big data analysis. The fundamental direction of systems engineering is unpacked around 12 chapters, which consider the process of evaluating the outcomes of the key parts outlined above. The chapters provide significant level of depth for the reader with an emphasis of providing a clear understanding of system reliability, system design analysis, simulation modelling, network management protocols, and business intelligence tools for decision-making processes.

Finally we consider the current challenges in the multidisciplinary field of strategic engineering namely the future direction of systems engineering and the way it is shaped to match and complement the global environment, the changing societal needs, the challenges faced by business and the key policy drivers as well as the technologies that these future systems undertake. The technological advances aligned with the basic fundamental components, their subsystems and infrastructure will no doubt create and increasing leap into the future leading to erudite services and products. The book is structured in such a way so as the readers can follow the book, chapter by chapter sequentially or they can 'dip into' the book chapters as they please.

The main emphasis of the book is the fundamentals of strategic engineering by outlining the trends on the ground rules for through-life systems with a view of addressing simulation modelling in line with the systems engineering constructs. The book introduces 12 chapters and presents interesting and insightful discussions in terms of the growth in the area of cloud and big data analytics, dealing with phenomena such as software process simulation modelling for agile cloud, the impact of business intelligence on organisations and strategic approaches to cloud computing. The individual chapters included in each part of the book are briefly summarised.

Chapter "Mathematical and Computational Modelling Frameworks for Integrated Sustainability Assessment (ISA)" focuses on outlining generic mathematical and computational approaches to solving nonlinear dynamical behaviour of complex systems. The goal of the chapter is to explain the modelling and simulation of system's responses experiencing interaction change or interruption (i.e., interactive disruption). Chapter "Sustainable Maintenance Strategy Under Uncertainty in the Lifetime Distribution of Deteriorating Assets" considers random variable model and stochastic Gamma process model as two well-known probabilistic models to present the uncertainty associated with the asset deterioration. Within Chapter "A Novel Safety Metric SM_{EP} for Performance Distribution Analysis in Software System" the focus is primarily on safety attributes becoming an essential practice towards the safety critical software system (SCSS) development. Chapter "Prior Elicitation and Evaluation of Imprecise Judgements for Bayesian Analysis of System Reliability" examines suitable ways of modelling the imprecision in the expert's probability assessments. Chapter "Early Detection of Software Reliability: A Design Analysis" takes the approach of design analysis for early detection of software reliability. Chapter "Using System Dynamics for Agile Cloud Systems Simulation Modelling" provides an in-depth background to cloud systems simulation modelling (CSSM) and its applicability in cloud software engineering— providing a case for the apt suitability of system dynamics in investigating cloud software projects. Chapter "Software Process Simulation Modelling for Agile Cloud Software Development Projects: Techniques and Applications" provides an overview of software process simulation modelling and addresses current issues as well as the motivation for its being—particularly related to agile cloud software projects. This chapter also discusses the techniques of implementation, as well as applications in solving real-world problems. Chapter "Adoption of a Legacy Network Management Protocol for Virtualisation" discusses, with examples, how network management principles could be contextualised with virtualisation on the cloud. In particular, the discussion will be centred on the application of simple network management protocol (SNMP) for gathering behavioural statistics from each virtualised entity. Chapter "Strategic Approaches to Cloud Computing" outlines strategic approaches to cloud computing with the focus on cloud providing business benefits when implemented in a strategic manner. Chapter "Cloud Security: A Security Management Perspective" focuses on strategic level, security considerations related to moving to the cloud. Chapter "An Overview of Cloud Forensics Strategy: Capabilities, Challenges and Opportunities" outlines a model for cloud forensics, which can be viewed as a strategic approach used by other stakeholders in the field, e.g., the court of law. Chapter "Business Intelligence Tools for Informed Decision-Making: An Overview" explains business intelligence and analytics concepts as a means to manage vast amounts of data, within complex business environments.

The objective of the book is to increase the awareness at all levels of the changing and enhanced technological environments we are living and working in, and how this technology is creating major opportunities, limitations and risks. The book provides a conceptual foundation, moving to a variety of different aspects of strategic engineering modelling approaches with the view of challenges not only faced by organisations but additional technological challenges we are consistently moving towards. Within this area we reflect upon the developments in and approaches to strategic engineering in a thematic and conceptual manner.

We hope that by introducing material on topics such as through-life sustainable systems, cloud computing, systems engineering, big data analytics systems modelling, we have been able to build knowledge and understanding for the reader; after reading this book the reader should be equipped with a greater appreciation and understanding concepts and the key alignment of strategic engineering within real-world case examples. There is only a limited amount which can be contained in each chapter; all of the chapter topics warrant a book in themselves. The focus is clearly on presenting a high-level view of relevant issues. We would further like to take this opportunity to thank the contributors for preparing their manuscripts on time and to an extremely high standard.

Leeds, UK   Amin Hosseinian-Far
                                                                                   Muthu Ramachandran
                                                                                   Dilshad Sarwar

Contents

Part I
Systems Lifecycle, Sustainability, Complexity, Safety and Security

Mathematical and Computational Modelling Frameworks for Integrated Sustainability Assessment (ISA)

Maryam Farsi, Amin Hosseinian-Far, Alireza Daneshkhah and Tabassom Sedighi

Abstract Sustaining and optimising complex systems are often challenging problems as such systems contain numerous variables that are interacting with each other in a nonlinear manner. Application of integrated sustainability principles in a complex system (e.g., the Earth's global climate, social organisations, Boeing's supply chain, automotive products and plants' operations, etc.) is also a challenging process. This is due to the interactions between numerous parameters such as economic, ecological, technological, environmental and social factors being required for the life assessment of such a system. Functionality and flexibility assessment of a complex system is a major factor for anticipating the systems' responses to changes and interruptions. This study outlines generic mathematical and computational approaches to solving the nonlinear dynamical behaviour of complex systems. The goal is to explain the modelling and simulation of system's responses experiencing interaction change or interruption (i.e., interactive disruption). Having this knowledge will allow the optimisation of systems' efficiency and would ultimately reduce the system's total costs. Although, many research works have studied integrated sustainability behaviour of complex systems, this study presents a generic mathematical and computational framework to explain the behaviour of the system following interactive changes and interruptions. Moreover, a dynamic adaptive response of the global

M. Farsi (✉)
Operations Excellence Institute, School of Aerospace, Transport and Manufacturing,
Cranfield University, Cranfield MK43 0AL, UK
e-mail: maryam.farsi@cranfield.ac.uk

A. Hosseinian-Far
School of Computing, Creative Technologies & Engineering, Leeds Beckett University,
Leeds LS6 3QR, UK
e-mail: A.Hosseinian-Far@leedsbeckett.ac.uk

A. Daneshkhah
The Warwick Centre for Predictive Modelling, School of Engineering,
The University of Warwick, Coventry CV4 7AL, UK
e-mail: A.Daneshkhah@warwick.ac.uk

T. Sedighi
Energy Environment Food Complexity Evaluation Complex Systems Research Centre,
Cranfield School of Management, Cranfield University, Cranfield MK43 0AL, UK
e-mail: t.sedighi@cranfield.ac.uk

© Springer International Publishing AG 2017
A. Hosseinian-Far et al. (eds.), *Strategic Engineering for Cloud Computing and Big Data Analytics*, DOI 10.1007/978-3-319-52491-7_1

3

system over time should be taken into account. This dynamic behaviour can capture the interactive behaviour of components and sub-systems within a complex global system. Such assessment would benefit many systems including information systems. Due to emergence and expansion of big data analytics and cloud computing systems, such life-cycle assessments can be considered as a strategic planning framework before implementation of such information systems.

1 Introduction

Sustainability can be defined as sustaining, preserving and enhancing some valuable or valued condition(s) over time in a dynamic system [12]. The sustainability science studies the complex relationship between nature and society in a global environment so-called global system. This complex interaction can occur between a broad range of sub-systems such as economic, ecological, technological, environmental and social notations [24, 35]. In the context of information systems, sustainability assessment is usually focused on the economy domain [33]. For instance, the interplay between human activities in a society and economy affects economic growth or decay, the standard of living, poverty, etc. Moreover, the interaction between human behaviour and ecological systems tends to focus on global warming, energy security, natural resources and biodiversity losses, etc. Finally, the interplay between humankind's actions, knowledge and activities and technological environment improve or regress technology, increase or decrease safety and has effects on the healthiness of people's daily lives. Meanwhile, Integrated Sustainability Assessment (ISA) applies sustainable principles to provide and support policies and regulations and incorporate decision-making in a global system across its life cycle. Therefore, ISA can be a solution-oriented discipline to evaluate the behaviour of a complex global system. The complete discipline integrates a broad range of knowledge and methodologies towards defining solutions. In this context, the development of a robust and integrated framework for sustainability assessment is of paramount importance.

Understanding the sustainability concept is clearly the basis for sustainability assessment. Sustainability or Sustainable development was first described by the Brundtland's report titled 'Our Common Future' published by the World Commission on Environment and Development. The paper argued that sustainable development means "development that fulfils the needs of the present without compromising the ability of future generations to meet their own needs. This contains two key concepts. (i) The concept of 'needs' in particular the necessary requirements of the world's poor, to which overriding priority should be given, and (ii) the idea of limitations imposed by the state of technology and social organisation on the environment's ability to meet present and future need" [13]. This argument tailed by several debates and discussions on how sustainability should be defined, interpreted and assessed [61]. Sustainability assessment can be described as a process to identify and evaluate the effects of possible initiatives on sustainability. The initiative can be a proposed or an existing policy, plan, programme, project, piece of legislation or a current practice or activity [63]. Developing transformative life cycle and system-oriented tools and methodologies both at national and international levels have been studied since late 1980s and earlier 1990s. Multiple factors and indicators can reflect the complexity of a global system's behaviour. This motivated efforts to combine indicators into integrated, quantitative measures of sustainability called Integrated Sustainability Assessment (ISA) [51].

In [5], Altieri investigated the critical issues which had an effect on productive and sustainable agriculture in Latin America using integrated pest management programs as case studies. He discussed that the attainment of such agriculture is dependent on new technological innovations, policy changes, and more socio-equitable economic schemes [5]. Furthermore, in [20] the concept of Triple Bottom Line (TBL) has been added to the accounting perception by John Elkington [20]. The TBL framework considers the interaction between three parts: social, environmental (or ecological) and economic in one global business system as illustrated in Fig. 1. Moreover, TBL introduced three dimensions commonly called the three Ps: People, Planet and Profit into the sustainability concept [21]. This new impression became a favourite subject in sustainability which requires consideration of economic, social and natural environmental parameters to characterise the valued conditions in sustainability [34]. Smith and McDonald [71] presented that agricultural sustainability assessment encompasses biophysical, economic and social factors. Therefore, they considered more parameters to assess the agricultural sustainability using multiple qualitative and quantitative indicators [71]. In the late 1990 s and early twentieth

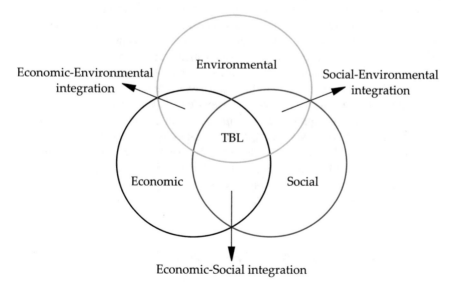

Fig. 1 The three pillars of sustainability or the three Ps: People, Planet and Profit; TBL is an integrated interaction between the environmental, economic and social dimension of global sustainability

century sustainability assessment has been developed by including the integrated sustainability science theory. In 1996, Rotmans and Asselt described integrated assessment as "integrated insights to decision makers" [65]. In this context, integration definition considers the combination of different parameters, functions, policies and regulations into a whole global system in an appropriate way that achieves a particular goal and purpose within the system. Integrated Sustainability Assessment (ISA) necessitates the following characteristics: (i) It should be conducted with explicit awareness of the global system; (ii) It should be directed by comprehensive aims, policies and regulations preferably explicitly articulated though initially less specifically; (iii) It usually requires stakeholder's involvement due to the significant influence of the policies and outcomes evaluation criteria on the direction of an integrated sustainability assessment; (iv) Sustainability assessment criteria are dynamic, time dependent and will change over time; (v) Integration is expensive since it necessitates a significant amount of time and different resources [12].

As mentioned earlier, assessment of integrated sustainability is a complex process due to the diversity of variables, parameters, formulations, policies and regulations in a global system. Sustainability assessment provides information for decision-makers to decide what action they should take or what policy they should apply within the system to make society more sustainable. The integrated assessment arises from Environmental Impact Assessment (EIA) and Strategic Environmental Assessment (SEA). However, SEA has been divided into economic and social approaches as well as the environmental aspect, which still reflects the Triple Bottom Line (TBL) approach to sustainability [29]. The overall aim of sustainability assessment is mainly to minimise unsustainability with respect to TBL objectives [11, 63]. Moreover, sustainability assessment can be conducted to evaluate a policy implementation into a system to inform, advise and update a management practice within the system. This study focusses on both mathematical and computational integrative methods within sustainability assessment. This multi-functionality and multiplicity of sustainability assessment terminologies and methodologies can be confusing. Therefore, this research study proposes a generic framework for the corresponding strengths of numerous approaches and methods which can be used by industrial ecologists and engineers and ecological and biophysical economists. This generic framework can also be used as starting point for many computer scientists and relevant industries for life-cycle assessments and applications of computing emerging technologies and systems such as big data analytics, systems set up, around cloud services and Internet of Things (IoT).

2 Integrated Sustainability Assessment (ISA)

This section discusses the triple pillars of sustainable development integration and describes the existing mathematical and computational methods concerning development of an integrated decision support system for sustainability assessment. In this context, the system is considered as a complex global system. As mentioned earlier, in a sustainable global system, the three pillars of sustainable development are Environmental, Economic (i.e., prosperity or profitability aspects) and Social. Initially, the characteristics and features of complexity in a system have been discussed by reviewing standard measures of complexity from relevant scientific literature. Then, in order to bring mathematical and computational flexibility and rigour to the issue, different statistical complexity measurements, and computational modelling features have been discussed.

A complex system is a system with numerous interdependencies in its structure. Such systems are sensitive to any small perturbations which may affect the initial conditions. In a complex system, the interactions between the independences and components are numerous. Therefore, the responses of the system to these perturbations and changes are not unique. Such behaviour is complex since the system can take multiple pathways to evolve. Moreover, the growth and development of a complex system may vary over time, and therefore this categorises such a system as a dynamic one. In this paper, the notion of complex dynamical system is denoted to a global system. A global system can be described as a System of Systems (SoS) which is composed of several sub-systems [33]. Analytically, the behaviour of such a complex dynamical system can be derived by employing differential equations or difference equations [77]. The key features and properties that can be associated with a global system are nonlinearity, feedback, spontaneous order, robustness and lack of central control, hierarchical organisation, numerosity and emergence, [50].

Although, nonlinearity is one of the features of a complex system, the behaviour of such system can also be linear where the corresponding parameters and objectives can be written as a linear sum of independent components. Meanwhile, linear studies can be used to understand the qualitative behaviour of general dynamical systems [36]. Analytically, this can be achieved by calculating the equilibrium points of the system and approximating it as a linear trend around each such point. Nonlinearity is often considered to be a vital characteristic of a complex system. The nonlinear system is a system that does not satisfy the superposition principle. The nonlinear behaviour of a global system becomes more complex regarding the diversity and variety of sub-systems' properties, conditions, and boundaries other than sub-systems variations. For instance, the complexity of living organism as a global system with numerous properties of being alive or dead other than a variety of its sub-systems and components (e.g., Human organisms composed of trillions of cells which are clustered into particular tissues and organs). Mathematically, for such global systems, typically, it is required to generate nonlinear differential equations to explain their dynamic behaviour.

Feedback is an important feature of a global system since the interaction between sub-systems are dynamic and are changing over time. Therefore, the behaviour of a nonlinear dynamic system extensively depends on the accuracy of the relations and interactions between its components. One of the typical examples of such systems in this context is the behaviour of a colony of ants that interact with each other. The quantitative approach to model such a complex behaviour can be defined by a nonlinear first-order differential equation, so-called System Dynamics (SD) which will be discussed later in this chapter. Since a complex dynamical system can be composed of a large number of elements, identifying the order of interaction between these elements are difficult and not clear. The spontaneous behaviour of orders is one of the most perplexing problems to define the behaviour of the complex systems on the feedback and information processes within the system over time. Moreover, the orders in a complex system can be robust in the system due to the scattered origin. Despite the sensitivity of a global system to any perturbation, orders are stable under such conditions. For example, consider a group of flying birds as a global system; they stay together, and the order of the pathways they take do not change, despite for instance any internal disruption such as the individual motion of the members or external disruption caused by the wind. Mathematically, robustness can be formulated in a computational language as the capability of a complex system to correct errors in its structure [68].

As mentioned earlier, a global complex dynamical system is composed of components and elements, so-called sub-systems. Such structures can resemble the structure of a hierarchical organisation. Subsequently, numerosity is an inherent property of a complex structure referring to the numerous number of sub-systems and parts in one global system. In a complex system, there may be different levels of organisations with individual properties and features. Emergence can arise from a robust order due to the complex interaction between these organisations and sub-systems within the global system. In system theory, emergence process between sub-systems is fundamental of integrative levels and complex systems [70].

Hitherto, different characteristics and properties of a global system have been explained briefly. Consequently, integrated assessment contributed to the sustainability science through developing a framework for different analytical and computational methodologies considering uncertainties, life-cycle thinking, policies, plans and regularities. In the following section, a different aspect of Integrated Sustainability Assessment (ISA) with regards to TBL is discussed, followed by introducing the related methodologies, models, tools and indicators. Moreover, ISA can be developed to define and evaluate the relationships between environmental, economic and social dimensions intended for optimising the interactive outcomes considering the TBL boundaries and constraints.

2.1 Environmental Sustainability Assessment

The environmental aspect of sustainability assessment is mainly viewed as ecological sustainability. The fundamental principles of environmental sustainability assessments focus on reducing the constructional embodied energy and CO_2 emissions; reducing the life cycle of atmospheric emissions (i.e., CO_2, NO_x, SO_x, CH_4, N_2O, etc.) and the waterborne emissions (i.e., COD, BOD, Total P, Total N, etc.) and limiting the requirement for water, fossil resources and natural gas [69]. Environmental sustainability assessments can provide indicators and indices as qualitative measurements to represent the environmental development in a defined system [35]. Indicators should be simple and transparent, measurable and quantifiable, sensitive to change and interruptions and time-dependent [58]. In particular, there are a number of environmental assessments developed by the Office for National Statistics (ONS). The UK's Office for National Statistics (ONS) is one of the largest independent national statistical institutes. One of the main tasks of ONS is to collect, develop and publish official statistics related to environment, economy, population, and society at national, regional and local levels [59]. According to the ONS, the headline 'environmental assessments' considers greenhouse gas emissions, natural resource use, wildlife: bird population indices and water use. Moreover, the corresponding supplementary measures are as follows: UK CO_2 emissions by sector, energy consumed in the UK from renewable sources, housing energy efficiency, waste, land use & development, origins of food consumed in the UK, river water quality, fish stocks, status of species & habitats and UK biodiversity impacts overseas [59].

The main and fundamental benchmark for evaluating the impact of environmental dimension on sustainability can be done by assessing the environmental performance of a global system using Environmental Life-Cycle Assessment (LCA). However, this assessment is not sufficient to understand the dynamic and interactive behaviour of environmental and ecological impacts. LCA is a simplified quantitative approach based on the traditional linear behaviour of complex systems. Despite the LCA's limitations and challenges in obtaining qualitative data, LCA can be still considered as the most comprehensive approach for environmental impact assessment [32]. The International Standards Organisation (ISO) is the institution that is responsible for establishing principles and guidelines for life-cycle assessment [76] called ISO 14040:2006– Principles and Framework [40] and ISO 14044:2006–Requirements and Guidelines [41] for LCA. According to the ISO standards, life-cycle assessment falls into two distinct classes based on its origin as Economic Input-Output based LCA (EIO) and Ecologically based LCA (Eco-LCA). Moreover, the corresponding LCA terms and expressions are defined as follows:

- Life cycle: "consecutive and interlinked stages of a product system, from raw material acquisition or generation of natural resources to final disposal".
- Life-cycle assessment (LCA): "compilation and evaluation of the inputs, outputs and the potential environmental impacts of a product system throughout its life-cycle".
- Life-cycle inventory analysis (LCI): "the assessment involving the compilation and quantification of inputs and outputs for a product throughout its lifecycle".
- Life-cycle impact assessment (LCIA): "the assessment aimed at understanding and evaluating the magnitude and significance of the potential environmental impacts of a product system throughout the lifecycle of the product".

Therefore, the life-cycle assessment is carried out in four distinct phases as illustrated in Fig. 2. These four phases are (i) Goal and scope definition, (ii) the Life-Cycle Inventory Analysis (LCI), (iii) the Life-Cycle Impact Assessment (LCIA) and (iv) the Life-Cycle Interpretation phase. Goal and scope phase focuses on identifying the aim of the LCA study and the corresponding objectives and applications with regards to the system boundaries, assumptions and constraints. Afterwards, in the inventory analysis phase, the essential data are collected to fulfil the objectives

Fig. 2 Life-Cycle
Assessment phases

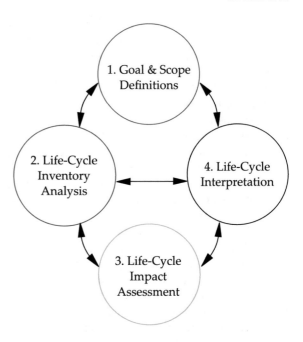

of the LCA study. The data collection can be achieved by inventorying the input and output data from the studied system. Subsequently, the results from inventory analysis are transformed into the corresponding environmental impacts such as utilisation of natural resources, human health and safety, etc. in the impact assessment phase. The final phase of the LCA study is life-cycle interpretation. In this phase, the results of the inventory and impact assessments are evaluated in order to provide information for decision-makers which will ultimately help to develop essential policies and recommendations based on the goal and scope definition in the studied system. Recently, ISO published a general guideline on the implementation of ISO 14004 [39] which applies to any organisation with respect to their environmental principles, products and delivered services.

Considering different life-cycle assessment approaches; Economic Input-Output Life-Cycle Assessment (EIO-LCA) method evaluates the required raw materials and energy resources, estimates the environmental emissions, and assesses the economic activities in a system. EIO-LCA is one of the LCA techniques for life-cycle assessment study which is helpful for evaluating environmental impacts of a product or a process over its life cycle. This method has been invented and developed by Wassily Leontief in [53]. Economic Input-Output (EIO) theory introduces a mathematical modelling approach to systems life-cycle assessment and therefore can be beneficial to industry. It can be used for evaluating the financial transactions based on inputs and outputs within the sector. EIO model can be formulated as a matrix X_{ij} which represents the financial transaction between sectors i and j in a particular year. Therefore, using this method, the necessity of the input for an internal transaction in a sector can be indicated as (i.e.,: $X_{ij} \neq 0$ where $i = j$). Moreover, the direct, indirect and total effects of alterations to the economy can be identified using linear algebra techniques [53].

Consider matrix A_{ij} as the normalised output amounts for sector j, so that $A_{ij} = X_{ij}/x_j$. In addition, consider a vector of final demand, y_i, for the output from sector i. Therefore, the total output from sector i, x_i can be calculated as the summation of the output from sector j as a consumer and the total transaction between other sectors X_{ij}, thus

$$
\begin{aligned}
x_i &= y_i + \sum_j X_{ij} \\
&= y_i + \sum_j A_{ij} x_j,
\end{aligned}
\tag{1}
$$

the vector notation of the Eq. 1 is thus:

$$
x = y + Ax \quad \Rightarrow \quad x = (I - A)^{-1} y.
\tag{2}
$$

Material Flow Analysis (MFA) is an analytical method for quantifying stocks and flows of materials and entities within a system. MFA is capable of tracking the entities (e.g., materials, productions, etc.) and evaluating their utilisation in a system. Combining traditional economic input-output modelling approach with material flow analysis creates a mixed-unit input-output analysis technique to track and evaluate the economic transactions under changes in productions [31]. The other method to estimate the direct and indirect resource requirements of a system is the well-known Physical and Monetary Input-Output (PMIO) method which evaluates all the physical flows associated with the system economy [81].

In addition to the methodologies and techniques discussed, in order to assess the economic aspect of the life cycle, the Ecologically based Life-Cycle Assessment (Eco-LCA) method evaluates the role of ecosystem services in life-cycle assessment. Eco-LCA classifies resources into the following categories: (i) Renewable versus non-renewable, (ii) biotic versus abiotic, (iii) materials versus energy or (iv) regarding their originating ecosphere (lithosphere, biosphere, hydrosphere, atmosphere and other services). Eco-LCA is a physically based approach to assess the flows and interactions between considered resources in a system. For instance, Material Flow Analysis (MFA), Substance Flow Analysis (SFA) and Energy Flow Analysis (EFA) are some physical flow-based assessments for Eco-LCA of a system. Material flow analysis focuses on biophysical aspects of human activity with a view to reducing environment-related losses. Substance flow analysis is an analytical method to quantify flows of certain chemical elements in an ecosystem. However, energy flow analysis focuses on the flow of all types of energy such as exergy and emergy-based on the first law of thermodynamics.

2.2 *Economic Sustainability Assessment*

The second dimension of sustainability assessment is Economic Sustainability Assessment (ESA) which is focused on estimation of prosperity and profitability of a global system over its life cycle [30]. ESA can be generated to estimate the required economic growth, necessary for maintaining the sustainability of a system. In particular, the economic dimension of sustainable development in the UK can be measured by the following benchmarks: economic prosperity, long-term unemployment, poverty and knowledge and skills measure. Moreover, the supplementary measures are as follows: population demographics, debt, pension provision, physical infrastructure, research and development and environmental goods & services sector [59]. One example of the economic indicators for monitoring economic prosperity is called Gross Domestic Product (GDP), which represents the scale of economic activities within the country. This indicator is developed by the UK Office of National Statistics (ONC). Similarly, Domestic Material Consumption (DMC) is an important indicator to measure resource productivity in the context of Europe 2020 strategy [22]. DMC indicator relates to the gross domestic products which are developed by Statistical Office of the European Communities (Eurostat). Eurostat is the statistical office of the European Union and is responsible for providing statistics, accounts and indicators supporting the development, implementation and monitoring of the EUs environmental policies, strategies and initiatives [73]. Moreover, Net National Product (NNP) is another economic indicator which represents the monetary value of finished goods and services. NNP can be calculated as the value of GDP minus depreciation. In accountancy, depreciation refers to the amount of GDP required to purchase new goods to maintain existing. GDP and NNP are the two most frequently used benchmarks by decision-makers, are defined to measure and assess the overall human welfare [58].

Life-Cycle Cost Analysis (LCCA) is the most powerful benchmark for evaluating the economic impact of sustainability. LCCA is applied to evaluate the economic performance of a system over its entire life. The cost assessment considers the total cost including the initial cost, operating and service cost, maintenance cost, etc. Therefore, identifying activities and subsequently, the associated costs is the initial requirement for conducting the life-cycle assessment study. Although, the Environmental Life-Cycle Assessment (ELCA) methodologies where the focus was on the material, energy and resources flow, within the Life-Cycle Cost Assessment (LCCA) money flows would be the main focus. Furthermore, the Full Cost Environmental Accounting (FCEA) is another life-cycle costing analysis method. Both LCCA and FCEA approaches assess the environmental cost, and therefore they are appropriate methods for evaluating the economic impacts of sustainability. Within the context of information systems, this pillar of sustainability is the major context where sustainability and resilience of the system are assessed. Hosseinian-Far and Chang outlined the metrics required for assessing the sustainability of selected information systems. It is also argued that such economic sustainability assessment of information systems and the use of metrics is context dependent [33].

2.3 Social Sustainability Assessment

The third pillar of sustainability assessment is referred to as Social Sustainability Assessment (SSA). This assessment provides measures and subsequent guidelines required for identifying social impacts on sustainability assessment in a global system. In this context, a global system is composed of numerous entities. These entities can be organisations, individuals, shareholders, stakeholders, etc. In such a system these entities have an obligation to provide services and actions for the benefit of society as a global system. This responsibility is called Social Responsibility (SR). In another word, SR is a duty every entity has to perform so as to maintain a balance between the economic and the ecosystems [8]. According to the UK Office for National Statistics (ONS), the following measures are the social benchmarks of sustainable developments: healthy life expectancy, social capital, and social mobility in adulthood and housing provision. Furthermore, supplementary social indicators are as follows: avoidable mortality, obesity, lifestyles, infant health, air quality, noise and fuel poverty.

Social Life-Cycle Assessment (SLCA) is the third dimension of life-cycle sustainability assessment which assesses the social and sociological impact of an organisation, individuals, and products along the life cycle [25]. In addition, the Millennium Ecosystem Assessment (MEA) estimates the impact of ecosystem changes on human well-being by considering health and security, social relations, freedom, etc. [8]. These social aspects are gathered and developed by the Sustainable Consumption and Production (SCP) in order to generate and implement the corresponding policies and actions for public and private decision-makers. However, identifying and evaluating social aspects are challenging since some of the social performance data are not easily quantifiable [66]. These aspects can be categorised as human rights, working condition, health and safety, cultural heritage, education, etc. [10].

Regarding the society impacts on sustainability assessment, the Joint United Nations Environment Programme (UNEP) and the Society of Environmental Toxicology and Chemistry (SETAC) develop knowledge and provide support in order to put life-cycle thinking into effective practices. This international life-cycle partnership is known as the Life-Cycle Initiative (LCI), established in 2002 [43]. SETAC is a worldwide and not for profit professional society that its main task is to support the development of principles and practices for support, improvement and management of sustainable environmental quality and ecosystem integrity [67]. Whereas, the UNEP as the leader of global environmental authority develops and implements the environmental dimension of sustainable development within the United Nations (UN) system and intends to support the global environment [10, 25].

Social assessment techniques and methodologies focus on improvement of social conditions since the main goal in SLCA is enhancing human well-being [9]. Human well-being can be evaluated through many indicators and terms. The most common ones are human development, standard of living and quality of life. For instance, Human Development Index (HDI) is an integrated indicator composed of life expectancy, education and income used by the United Nations Development

Programme (UNDP). This indicator intends to evaluate the combined social and economic growth in a system [56]. Moreover, the Physical Quality of Life Index (PQLI) is the other SLCA indicator which is composed of life expectancy at age one, infant mortality and literacy rate developed by David Morris in the mid-1970s at the Overseas Development Council (ODC) [23].

3 Integrated Mathematical and Computational Methodologies

Mathematical and computational methodologies used for evaluating the integrated sustainability assessment (ISA) can fall into two main categories: (i) Computing and Information Science (CIS) and (ii) Integrated systems modelling. CIS focuses on both computing and informatics aspects facilitating ISA by analytical systems modelling such as data mining & analysis, artificial intelligence and dynamical simulation. However, the integrated systems modelling develops computational and quantitative sustainability assessment. Meanwhile, the computational aspect focuses on web-based databases, cloud computing cyberinfrastructure, advanced data acquisition and artificial intelligence. Though, the quantitative aspect has also implemented sustainability assessment within operations research and management science [79].

The main component of integrated sustainability assessment is the life-cycle techniques themselves [25]. Application of life-cycle principles is essential to achieve reliable sustainability assessment results. The life-cycle principles consider all types of life cycles such as environmental, economy and social for all categories of subsystems in a global system. As mentioned earlier, these sub-systems can be products and materials, organisations and stakeholders, supply chains, manufacturing processes, etc. There are several mathematical and computational techniques to estimate the integrated sustainability assessments composed of environmental, economic and social impacts. Hitherto, different terms and indicators of the three pillars in sustainability assessment have been explained and discussed. The initial stage of Integrated Sustainability Assessment (ISA) is to identify and evaluate the critical corresponding sustainability criteria, indicators and metrics. This can be estimated by using Pareto principle based on 80–20 rule. The 80–20 rule is also known as the law of the vital few states in which 80% of the effects arise from 20% of the causes approximately [6]. Further to this estimation, different mathematical and computational methodologies for ISA are introduced and explained briefly in the following sections. These integrated methodologies are: Klöffer technique, Multi-criteria decision analysis (MCDA) & Multi-objective decision-making (MODM), System Dynamics (SD), Agent-Based Modelling (ABM), Sustainability Network Analysis and Sustainability Optimization and Scenario Analysis.

3.1 Klöepffer Technique

Klöepffer [49] proposed that the integrated life cycle in a sustainability assessment can be simply calculated by the summation of environmental, economic and social life-cycle assessment calculations, thus

$$LCSA = ELCA + LCCA + SLCA \tag{3}$$

Although the Klöffer technique considers the triple pillars of sustainable development, it does not reflect on the complexity of interaction between these three dimensions. This is due to the linear equation with equal weighted components in the Klöffer method. To tackle this problem, different mathematical and computational methods have been developed which will be discussed in the following sections. Meanwhile, for instance, in a more recent computational method, the open source OpenLCA software offers an integrated and comprehensive method for analysing life-cycle impact assessment. This method considers different impact categories including normalisation and weighting factors [60]. The software includes features such as graphical modelling & simulation and uncertainty approximation.

3.2 Multi-criteria Decision Analysis (MCDA)

Achieving an integrated sustainability necessitate an integrated decision-making. Decision problems consist of making choice(s), generating ranking(s) and sorting problem(s). These problems are often complex since they involve several criteria and aspects. Therefore, to create a sustainable decision, it is required to consider multiple criteria in the decision process so-called Multi-criteria decision analysis (MCDA) or Multi-objective decision-making (MODM) analysis [38]. There are numerous methods to perform MCDA or MODA in a decision problem. Some of these methods include: Analytic Hierarchy Process (AHP), Analytical Network Process (ANP), Multi-Attribute Utility Theory (MAUT/UTA), etc. In general, all MCDA methods are based on the following structure in order to develop an integrated and sustainable decision-making. This structure involves as thus: (i) Criteria selection; to select n sustainability criteria including technical, environmental, economic, and social criterion. (ii) Alternative selection; to select m local (with respect to one specific criterion) and global (with respect to all criteria) alternatives. (iii) Grouped decision matrix development; where element x_{ij} is the performance of j-th criteria C of i-th alternative A. Besides, each criterion c_j is weighted by a positive weight w_j calculated as thus

$$
\begin{aligned}
C &= \begin{bmatrix} c_1 \ c_2 \ c_3 \ \ldots \ c_n \end{bmatrix}, \\
W &= \begin{bmatrix} w_1 \ w_2 \ w_3 \ \ldots \ w_n \end{bmatrix}, \\
A &= \begin{bmatrix} a_1 \ a_2 \ a_3 \ \ldots \ a_m \end{bmatrix}.
\end{aligned}
\tag{4}
$$

Moreover, a positive weight is assigned to a criterion to specify its relative importance among all criteria. The attributed weight arises from the variance and the independency degree of criteria, and the subjective preference of the decision-maker(s) [80]. There are two main methods to calculate the weighted criteria: equal weights method and rank-order weights method. In equal weights method, w_j is calculated as below

$$w_j = \frac{1}{n}, \qquad j = 1, 2, , n. \tag{5}$$

Since the equal weights method does not consider the interactions between criteria, rank-order weighting method is proposed with criteria weights distribution as

$$w_1 \geq w_2 \geq w_3 \geq\geq w_n \geq 0 \quad where, \sum_{i=1}^{n} w_i = 1. \tag{6}$$

Furthermore, the rank-order weighting method is categorised into three different methods: (1) subjective weighting method (e.g., pair-wise comparison, analytical hierarchy process (AHP), etc.), (2) objective weighting method (e.g., entropy method, technique for order preference by similarity to ideal solution (TOPSIS), etc.) and (3) combination weighting method (e.g., ANP, MAUT) [48]. The weight of a criteria ranked to be jth is calculated as [80, 87]

$$
\begin{aligned}
(1) \quad & w_{1j} = \frac{1}{n} \sum_{k=i}^{n} \frac{1}{k} \\
(2) \quad & w_{2j} = \frac{1 - H_j}{n - \sum_{j=1}^{n} H_j}, \quad where: \quad H_j = -\frac{1}{\ln m} \sum_{i=1}^{m} \frac{f_{ij}}{\ln f_{ij}}, \\
(3) \quad & w_{3j} = \frac{w_{1j} w_{2j}}{\sum_{j=1}^{n} w_{1j} w_{2j}},
\end{aligned}
\tag{7}
$$

where, f_{ij} is the joint probability density function, k is the linear combination coefficient and $k \geq 0$ and w_{1j} and w_{2j} are subjective and objective weights respectively.

MCDA methods can be applied to develop different analysis such as (i) Stakeholders Analysis, (ii) Sensitivity and Uncertainty and Variability Analyses (e.g., Perturbation Analysis), (iii) Geographic Information Systems (GIS) and (iv) Data Envelopment Analysis (DEA). However, as mentioned earlier, to make the results more expedient for decision and policy makers, it is required to consider the uncertainty and dynamic interrelationships and interactions between the variables over time. In this regard, a number of appropriate methods are: Dynamic Multiple Criteria Decision Analysis (DMCDA), Stochastic Multi-Criteria Decision-Making (SMCDM) [14, 37] and fuzzy-MCDA [15, 44], System Dynamics (SD), Agent-Based Modelling (ABM) and Network. In the following, some of these methods are explained briefly.

As an extension of static MCDA problem, Dynamic-MCDA (DMCDM) considers variables over time across several criteria [14]. Therefore, a set of time periods as matrix $T = [t_1, t_2, \ldots t_k \ldots t_\tau]$ will be added to the Eqs. (4–7) and therefore $x_{ij}(t_\tau)$ represents the performance of criterion c_j for alternative a_i at t_τ time period. Therefore, in DMCDM, it is also required to collect alternative data over different time periods and criteria. Subsequently, aggregate value data $R = [r_1, r_2, \ldots r_k \ldots r_\tau]$ from different time periods can be calculated. Finally Data Envelopment Analysis (DEA) based results can be obtained by maximising the possibility of alternative $V(a_i)$ as [14]

$$V(a_i) = \max_{\varepsilon \leq r_k \leq 1} V(a_i) = \sum_{j=1}^{n} w_j . v_j^i, \tag{8}$$

where, v_j^i is the overall value of alternative a_i over criterion, c_j over τ time periods and expressed as thus,

$$v_j^i = \sum_{k=1}^{\tau} r_k . v_j^i(t_k), \tag{9}$$

where $v_j^i(t_k)$ is the value of alternative a_i on criterion c_j at t_k time period. Furthermore, stochastic MCDA is the other extension for static MCDA which includes probabilistic distribution for the decision alternatives across the decision criteria [72]. This distribution reflects possible uncertainties due to the incomplete data. Moreover, fuzzy-MCDA is the other mathematical tool in the decision-making process under uncertain conditions using fuzzy sets [85]. Although, fuzzy-MCDA is based on stochastic models, in this technique, decision alternatives uncertainties can be associated with fuzziness concerning the criterion weight assessment [55].

3.3 System Dynamics (SD)

The world as a global and complex system is constantly changing. Despite all available powerful mathematical and analytical methods and tools concerning sustainability, the interrelationship between different aspects of such a complex system necessitates a number of decision-makings and policy analysis. System Dynamics (SD) as a method, provides a networking technique to fulfil this concern in complex systems. The networking feature in SD method reflects the complex interactive behaviour of all systems components considering consequences of each action and decisions through the whole system. This is the main and fundamental aspect of SD that makes this method very powerful to solve complex systems compared to other methods. Long-term effective policies and decisions for sustaining a complex system are vital. Therefore, considering the frequent feedback of all components, variables and indicators over time are key to decision-making. System Dynamics creates a feedback loop between the decision made and the state of the system after

Fig. 3 System Dynamics
(SD): Stock and Flow
diagram; the sign of the links
represents whether the
variables at the two end
move in the same (+) or
opposite (−) directions; the
sign for the loop represents
whether it is a positive (+) or
negative (−) feedback loop
[47]

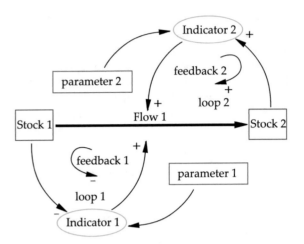

implementation of such decisions [74]. This assists the decision-maker in optimising the decision with minimum negative consequences (or side effects) within the
system. Although an SD network intends to expand boundaries of global systems
to consider the impacts of sub-systems on each other, there might be limits in the
decision-maker's understanding.

The global network of a complex system in the system dynamics method com-
poses of three main parts, "stock" are accumulations of sub-system (i.e., agents in
ABM method), "flow" as the rate of change between stocks within the network and
"delay", which is the time-delayed interrelationship between the system quantitative
measurements [26] as illustrated in Fig. 3. Delays can be caused by a variety of reasons such as inertia, delays in communication or any barrier that might delay the
message passing between the entities. Regarding integrated sustainability, the three
pillars of sustainability: environmental, economic and social can be modelled using
the network configuration in SD. Moreover, this modelling technique is capable of
performing multiple scenarios. This comparison capability would assist a decision-
maker or a policy developer to compare different scenarios with a view to selecting
the favourable solution. Therefore, combination of system dynamics and scenario
analysis (which will be discussed later in this chapter), is a dominant methodology
for decision-making and subsequently optimisation of complex systems. Mathematically, the quantitative approach to model a complex system's sustainment using SD
can be defined by a nonlinear first-order differential equation as,

$$\frac{d}{dt}x(t) = A.f(x, p_i), \tag{10}$$

where x is the state vector at time t, dt is the discrete intervals of systems time length,
A is the constant matrix, p_i is a set of parameters, f is the nonlinear smooth evolution
function affects the system's sustainability. Moreover, regarding the computational
approach, there are a number of appropriate computer software packages to facilitate

the development, modelling and simulation of SD models such as STELLA, iThink, Powersim and Vensim and AnyLogic [30].

Ahmad and Simonovic [2] combined system dynamic approach and Geographic Information System (GIS) so-called Spatial System Dynamics (SSD) to study flood management in the Red River basin in Manitoba, Canada. The proposed approach considers the interaction between the components of the studied system using a feedback-based modelling through the dynamic behaviour of processes over time [2]. Moreover, Videira et al. [79] discussed the application of system dynamics approach on the integrated sustainability assessment framework to fulfil policy making in complex systems. The proposed framework intended to provide an improved insight into the dynamical behaviour of complex systems regarding long-term sustainability impacts on decision-making [79]. Furthermore, Xu [83] studied an integrated sustainability assessment of urban residential development using System Dynamics (SD) methodology. To explain the interaction between local and global aspects of the urban residential development system, SD has been combined with GIS approach. Consequently, the proposed method provide a better and comprehensive insight for decision-maker [83]. This can also be referred to as Urban Dynamic in some textbooks. Similarly, the dynamic behaviour of Triple Bottom Line (TBL) of sustainability in interaction with each other has been studied by Lee et al. [52] using the system dynamic method. The proposed approach studied the dynamical and multidimensional characteristics of a product service system with a view to assessing the integrated sustainability of the system through a comprehensive approach [52]. In a more recent study, Abadi et al. [1] investigated the sustainability of a water resources management systems using system dynamics approach developed in Vensim software package. The study focuses on scenario analysis based on Analytical Hierarchy Process (AHP) for prioritising the sustainability indicators within the simulated scenarios. The results proposed an extensive insight for policy and decision-makers in the management system of water resources [1].

3.4 Agent-Based Modelling (ABM)

Agent-Based Modelling (ABM) is a computational method for simulating complex systems considering dynamic behaviour and nonlinear interactions between multiple components so-called agents. Although even today, the significance and decency in the majority of science is still based on mathematics, formulas, symbols and Greek letters, formulating most of the real and complex problems is extremely challenging. Therefore, computational methods offer this opportunity to study many of these complex phenomena in a system. In this regards, ABM is relatively a new approach to developing an integrated sustainability assessment in a complex system. Agents and sub-systems are introduced to the computational program with their corresponding metrics, properties and indicators. The overall aim of the simulation is to obtain the global consequences of the individual and interactive behaviour of all sub-systems in a given geographical area over a specific period. In contrast with the mathematical

Fig. 4 Agent-Based modelling (ABM) presented the triple pillars of integrated sustainability assessment in a global system

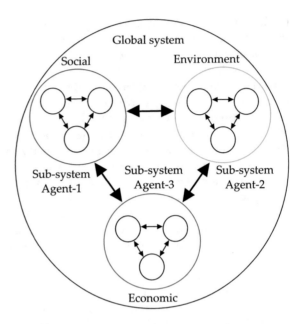

methods which represent components (the "agents") by variables, ABM introduces all agents into the modelling space based on their behaviours and characteristics. However, agent-based modelling use "what if" and "if-the" scenarios frequently to reflect the complex behaviour of a system which is clearly based on a mathematical formula. Therefore, agent-based models consist of agents (e.g., local or sub-systems), specific environment and domain (e.g., global system), and interaction rules between different agents [7].

Regarding life-cycle sustainability assessment, ABM technique is capable of combining the three pillars of sustainability in order to provide an integrated information for the decision-maker as presented in Fig. 4. ABM provides a nonlinear dynamic system to represent the behaviour of a complex global system with or without uncertainties within the sub-systems. Compared to the system dynamics method, ABM is more flexible for modelling a complex system with variable and multiple characteristics in sub-systems' behaviour and their interactions' structure [19]. There is a wide range of ABM applications in integrated sustainability assessment concerning sustainable mobility [82], policy analysis in decision-making [75], integrated climate change modelling [57], integrated land system [27], etc. Agent-based modelling and simulation packages and software are based on Object-Oriented Programming (OOP) since each agent can be introduced as an object within the programming language. In OOP languages such as Simula, Java, Python, C++, NetLogo, etc. the focus is more on defining the logic when objects interact with each other in a complex system, than naming the entities and the programming itself. Some of the agent-based software packages are AnyLogic, MASON, MASS and Swarm.

3.5 Sustainability Network Analysis

In a complex system, the interactive behaviour of sub-systems (i.e., agents in ABM or stocks in SD methods) can be modelled as a graph comprising of discrete elements. In such a graph, the complex interactions between the elements can be represented as a network [28]. This graph-based network can illustrate the complex behaviour of a system in a more straightforward manner, and can provide an appropriate tool to solve such problems using the network theory [3]. Sustainability Network Theory (SNT) aims to reflect the interaction between the three pillars of sustainability: environmental, economic and social. In this context, Network Analysis (NA) is introduced briefly as a powerful tool for assessing integrated sustainability within a complex system.

The initial step for using NA to assess sustainability in a global system is identifying and defining data on all elements, sub-systems and parameters (i.e., agents in ABM or stocks in SD methods) using the simple input–output matrices. Subsequently, different corresponding indices are required to be calculated in order to measure the performance of the global system with regards to integrated sustainability assessment aspects. The method necessitates considering elements' constraints and boundaries individually and also when they are interacting with each other [62]. The output from network analysis provides information for decision-makers. The applications for such a method includes assessment of energy consumption, environmental emissions, the total cost of the global system, etc. Performing integrated sustainability assessment as a network structure is key to studying the complex behaviour and consequently the welfare of for instance a society as a global system. This complex behaviour arises from mechanical and strategic interplays and flows. Concerning mechanical aspect, network structure mainly represents the system's behaviour as a linkage and/or a transmission path between different elements and agents. This connection can be modelled probabilistically within network. Whereas, the complexity of network is to be modelled through its structure and interactive outcomes, a more complicated techniques such as dynamic and/or equilibrium analysis are required [42].

Prell et al. have studied the sustainability of upland management in Peak District National Park in the UK using combine social network analysis and stakeholder analysis. They argued that the combined method would improve stakeholders' representation in the participating processes. Moreover, they concluded that the analysis provides more detailed information, despite the increase in time and costs [64]. In a similar study, Lim et al. advanced the Prell's study by developing a software tool that identifies and prioritises stakeholders using social networks so-called StakeNet [54]. The sustainability assessment of socio-ecological system has been discussed within Gonzaleś et al. research in which network analysis metrics are used. This study initially focused on the resilience of the system as one of the sustainability characteristics followed by the robustness feature using quantitative network analysis metrics [28]. In a more recent study, integrated network theory has been applied to generate an environmental life-cycle assessment in the Chilean electricity energy system [46]. Network theory as a framework for studying integrated sustainability assessment is still in its early stages of development. Since the network analysis is based on network configurations and complex dynamic analysis algorithms, an integrated methodology for network model development is still required.

3.6 Sustainability Optimization and Scenario Analysis

Scenario analysis has been widely used in different applications for the past decades for optimisation studies, decision-making, policy planning and risk and uncertainty assessment. Scenario analysis begins with identifying indicators and metrics which define the current and the future states of a system. The next step is to identify uncertainties and risks within the system. These factors are required to be prioritised and weighted and therefore different scenarios will be created. The three main sources of uncertainty in the integrated sustainability assessment are: (i) sustainable development and the corresponding physical, economic and social boundaries, (ii) the inherent subjectivity of assessment tools and (iii) the imperfection of the modelling tools and incomplete data to mimic the real situation [17]. Since there is an inevitable connection between integrated sustainability assessment and risk (or uncertainty in this context), scenario analysis is an appropriate method for sustainability optimisation.

Optimisation and risk (or uncertainty) analysis for less complex systems can be obtained by linear programming methods such as sensitivity analysis. Sensitivity analysis can be applied as an extension to the integrated sustainability assessment's methods as discussed in the previous sections. However, for complex systems, stochastic programming methods such as Markov Chains, Markov Decision Process and Bayesian Network are more sensible to be applied in order to obtain the optimum behaviour of such systems. The stochastic programming methodology allows one problem to consider several scenarios simultaneously while scenarios present different consequences of random variables [84]. This can provide an appropriate and comprehensive policy information for the decision-maker.

Application of scenario analysis for studying integrated sustainability of an industrial sector has been studied by Allwood et al. [4]. They considered a wide-scale change to the clothing sector in the UK in order to predict the behaviour of this manufacturing sector in the future. Their results have been validated through extensive stakeholder discussions [4]. Moreover, Cinar and Kayakutlu [16] developed scenario analysis by using Bayesian Networks for decision-making and policy planning in an energy sector [16]. Their proposed method assisted the decision-makers in the studied energy sector by providing new policies.

4 Resilience

Moreover, when a complex system is disrupted by an event, the system emerges a property called "resilience" [78]. The disruptive event is exposed to all the boundaries of the system domain; therefore loss of resilience can be described as an unstable dynamic response. The definition of resilience delivers a comprehensive perspective on performance which is the capability of a system to maintain its sustainable (or stable) behaviour. Meanwhile, measuring the interactive resilience (arises from interactive disruption) of a complex system is becoming more challenging. The interactive resilience in an interrupted complex system can be modelled mathematically and computationally using Agent-Based Modelling (ABM) approach (Phillips, 2010). This can be possible by using the Hill function [7] which is a mathematically efficient method to find the stationary points of a system. However, due to the inherent complexity of input and output variables in complex systems, the ABM approach is becoming intractable, computationally time-consuming and complicated. Therefore, such a problem can be tackled by combining ABM with the Gaussian Process emulator. This combined methodology would create a computationally fast and efficient probabilistic approximation model [86]. Furthermore, the probabilistic prediction results can be used directly to generate risk analysis [18], model calibration [45], optimisation, forecasting the future, etc. without re-evaluating the ABM at any additional data points. Finally, the combined methodology is capable of explaining the behaviour of the complex system that experiences changes and interruptions.

5 Conclusions, Discussions and Further Work

Optimisation of complex systems is still a challenging problem as it contains numerous parameters and variables that are interacting with each other in a nonlinear manner. Functionality and flexibility assessment of a complex system is a key element for anticipating the systems' responses to changes and interruptions. This study discussed mathematical, and computational approaches for integrated sustainability assessment focused on solving the nonlinear dynamical behaviour of complex systems. Moreover, in conclusion, having this knowledge will allow the optimisation of systems' efficiency and would ultimately reduce the system's total costs. As discussed earlier, a complex system can be considered as a System of Systems [33] and its behaviour can be derived using mathematical modelling by employing differential equations or difference equations [77]. Understanding resilience and sustainability of a system would assist the scientists, engineers, managers and in general, all the stakeholders to consider the life cycle, benefits and drawbacks and the future of the system under development. This is also not an exception in information systems development. Cloud services and big data analytics are the emerging technologies within the computing and informatics discipline. Developing such information systems therefore, would require a thorough assessment and strategic view on the information systems' sustainability. The resilience of such a systems can also be studied for systems' safety and security assessment. In this chapter, we intended to offer a generic view on systems' sustainability and resilience and different approaches for their modelling, which can also be applied by the computer scientists when developing emerging IT systems.

References

1. Abadi, L., Shamsai, A., & Goharnejad, H. (2015). An analysis of the sustainability of basin water resources using Vensim model. *KSCE Journal of Civil Engineering, 19*(6), 1941–1949.
2. Ahmad, S., & Simonovic, S. (2004). Spatial system dynamics: new approach for simulation of water resources systems. *Journal of Computing in Civil Engineering, 18*(4), 331–340.
3. Ahuja, R., Magnanti, T., & Orlin, J. (1993). *Network flows: theory, algorithms, and applications.* Prentice Hall.
4. Allwood, J., Laursen, S., Russell, S., de Rodríguez, C., & Bocken, N. (2008). An approach to scenario analysis of the sustainability of an industrial sector applied to clothing and textiles in the UK. *Journal of Cleaner Production, 16*(12), 1234–1246.
5. Altieri, M. A. (1992). Sustainable agriculture sustainable agricultural development in Latin America: exploring the possibilities. *Agriculture, Ecosystems & Environment, 39*(1), 1–21.
6. Azapagic, A. (1999). Life cycle assessment and its application to process selection, design and optimisation. *Chemical Engineering Journal, 73*(1), 1–21.
7. Barnes, D., & Chu, D. (2010). Agent-based modeling. In *Introduction to modeling for biosciences* (pp. 15–77). Springer.
8. Benoît, C., Norris, G., Valdivia, S., Ciroth, A., Moberg, A., Bos, U., et al. (2010a). The guidelines for social life cycle assessment of products: just in time!. *The International Journal of Life Cycle Assessment, 15*(2), 156–163.
9. Benoît, C., Traverso, M., Valdivia, S., Vickery-Niederman, G., Franze, J., Azuero, L., et al. (2013). The methodological sheets for sub-categories in social life cycle assessment (S-LCA). In *United Nations Environment Programme (UNEP) and Society for Environmental Toxicology and Chemistry (SETAC).*
10. Benoît, C., & Vickery-Niederman, G. (2010). Social sustainability assessment literature review. *The Sustainability Consortium.*
11. Bertrand, N., Jones, L., Hasler, B., Omodei-Zorini, L., Petit, S., & Contini, C. (2008). Limits and targets for a regional sustainability assessment: an interdisciplinary exploration of the threshold concept. In *Sustainability impact assessment of land use changes* (pp. 405–424). Springer.
12. Brinsmead, T. S. (2005). *Integrated sustainability assessment: identifying methodological options* (Tech. Rep.). Australia: Joint Academies Committee on Sustainability of the National Academies Forum.
13. Brundtland, G. (1987). *Our common future: Report of the 1987 World Commission on Environment and Development* (pp. 1–59). Oslo: United Nations.
14. Chen, Y., Li, K., & He, S. (2010). Dynamic multiple criteria decision analysis with application in emergency management assessment. In *2010 IEEE International Conference on Systems Man and Cybernetics (SMC)* (pp. 3513–3517).
15. Chiou, H., Tzeng, G., & Cheng, D. (2005). Evaluating sustainable fishing development strategies using fuzzy MCDM approach. *Omega, 33*(3), 223–234.
16. Cinar, D., & Kayakutlu, G. (2010). Scenario analysis using Bayesian networks: a case study in energy sector. *Knowledge-Based Systems, 23*(3), 267–276.
17. Ciuffo, B., Miola, A., Punzo, V., & Sala, S. (2012). Dealing with uncertainty in sustainability assessment. *Report on the application of different sensitivity analysis techniques to field specific simulation models. EUR, 25166.*
18. Daneshkhah, A., & Bedford, T. (2013). Probabilistic sensitivity analysis of system availability using Gaussian processes. *Reliability Engineering & System Safety, 112*, 82–93.
19. Davis, C., Nikolić, I., & Dijkema, G. P. (2009). Integration of life cycle assessment into agent-based modeling. *Journal of Industrial Ecology, 13*(2), 306–325.
20. Elkington, J. (1994). Towards the sustainable corporation: Win-win-win business strategies for sustainable development. *California Management Review, 36*(2), 90–100.
21. Elkington, J. (2004). Enter the triple bottom line. *The triple bottom line: Does it all add up* (Vol. 11, issue no. 12, pp. 1–16).

22. European Union. (2014, 26 March). Environmental statistics and accounts [Computer software manual].
23. Ferrans, C., & Powers, M. (1985). Quality of life index: Development and psychometric properties. *Advances in Nursing Science, 8*(1), 15–24.
24. Fiksel, J. (2003). Designing resilient, sustainable systems. *Environmental Science & Technology, 37*(23), 5330–5339.
25. Finkbeiner, M., Schau, E., Lehmann, A., & Traverso, M. (2010). Towards life cycle sustainability assessment. *Sustainability, 2*(10), 3309–3322.
26. Forrester, J. W. (1997). Industrial dynamics. *Journal of the Operational Research Society, 48*(10), 1037–1041.
27. Gaube, V., Kaiser, C., Wildenberg, M., Adensam, H., Fleissner, P., Kobler, J., & Smetschka, B. (2009). Combining agent-based and stock-flow modelling approaches in a participative analysis of the integrated land system in Reichraming, Austria. *Landscape Ecology, 24*(9), 1149–1165.
28. Gonzales, R., & Parrott, L. (2012). Network theory in the assessment of the sustainability of social-ecological systems. *Geography Compass, 6*(2), 76–88.
29. Hacking, T., & Guthrie, P. (2008). A framework for clarifying the meaning of Triple bottom-line, integrated, and sustainability assessment. *Environmental Impact Assessment Review, 28*(2), 73–89.
30. Halog, A., & Manik, Y. (2011). Advancing integrated systems modelling framework for life cycle sustainability assessment. *Sustainability, 3*(2), 469–499.
31. Hawkins, T., Hendrickson, C., Higgins, C., Matthews, H., & Suh, S. (2007). A mixed-unit input-output model for environmental life-cycle assessment and material flow analysis. *Environmental Science & Technology, 41*(3), 1024–1031.
32. Heijungs, R., Huppes, G., & Guinée, J. (2010). Life cycle assessment and sustainability analysis of products, materials and technologies. Toward a scientific framework for sustainability life cycle analysis. *Polymer Degradation and Stability, 95*(3), 422–428.
33. Hosseinian-Far, A., & Chang, V. (2015a). Sustainability of strategic information systems in emergent vs. prescriptive strategic management. *International Journal of Organizational and Collective Intelligence, 5*(4).
34. Hosseinian-Far, A., & Jahankhani, H. (2015b). Quantitative and systemic methods for modeling sustainability. In M. Dastbaz, C. Pattinson, & B. Akhgar (Eds.), *Green information technology: A sustainable approach* (Chap. 5). UK: Elsevier Science.
35. Hosseinian-Far, A., Pimenidis, E., Jahankhani, H., & Wijeyesekera, D. (2010). A review on sustainability models. In *International Conference on Global Security, Safety, and Sustainability* (pp. 216–222).
36. Hosseinian-Far, A., Pimenidis, E., Jahankhani, H., & Wijeyesekera, D. (2011). Financial Assessment of London Plan Policy 4A. 2 by probabilistic inference and influence diagrams. In *Artificial intelligence applications and innovations* (pp. 51–60). Springer.
37. Hu, J., & Yang, L. (2011). Dynamic stochastic multi-criteria decision making method based on cumulative prospect theory and set pair analysis. *Systems Engineering Procedia, 1*, 432–439.
38. Ishizaka, A., & Nemery, P. (2013). *Multi-criteria decision analysis: Methods and software.* Wiley.
39. ISO 14004:2016. (2016, March). *Environmental management systems – General guidelines on implementation* (Standard). 1214 Vernier, Geneva, Switzerland: International Organization for Standardization.
40. ISO 14040:2006. (2006, July). *Environmental management—Life cycle assessment—Principles and framework* (Vol. 2006; Standard). 1214 Vernier, Geneva, Switzerland: International Organization for Standardization.
41. ISO 14044:2006. (2006, July). *Environmental management—Life cycle assessment—Requirements and guidelines* (Standard). 1214 Vernier, Geneva, Switzerland: International Organization for Standardization.
42. Jackson, M. (2010). An overview of social networks and economic applications. *The handbook of social economics* (Vol. 1, pp. 511–585).

43. Jolliet, O., Margni, M., Charles, R., Humbert, S., Payet, J., & Rebitzer, G., et al. (2003). IMPACT 2002+: a new life cycle impact assessment methodology. *The International Journal of Life Cycle Assessment, 8*(6), 324–330.
44. Jovanovic, M., Afgan, N., & Bakic, V. (2010). An analytical method for the measurement of energy system sustainability in urban areas. *Energy, 35*(9), 3909–3920.
45. Kennedy, M., & O'Hagan, A. (2001). Bayesian calibration of computer models. *Journal of the Royal Statistical Society: Series B (Statistical Methodology), 63*(3), 425–464.
46. Kim, H., & Holme, P. (2015). Network theory integrated life cycle assessment for an electric power system. *Sustainability, 7*(8), 10961–10975.
47. Kirkwood, C. (1998). System dynamics methods. *College of Business Arizona State University USA*.
48. Kirkwood, C., & Sarin, R. (1985). Ranking with partial information: A method and an application. *Operations Research, 33*(1), 38–48.
49. Klöepffer, W. (2008). Life cycle sustainability assessment of products. *The International Journal of Life Cycle Assessment, 13*(2), 89–95.
50. Ladyman, J., Lambert, J., & Wiesner, K. (2013). What is a complex system? *European Journal for Philosophy of Science, 3*(1), 33–67.
51. Lal, R., Ghuman, B., & Shearer, W. (1990). Sustainability of different agricultural production systems for a rainforest zone of southern Nigeria. In *Transactions 14th International Congress of Soil Science, Kyoto, Japan, August 1990* (Vol. vi, pp. 186–191) .
52. Lee, S., Geum, Y., Lee, H., & Park, Y. (2012). Dynamic and multidimensional measurement of Product-Service System (PSS) sustainability: a Triple Bottom Line (TBL)-based system dynamics approach. *Journal of Cleaner Production, 32*, 173–182.
53. Leontief, W. (1970). Environmental repercussions and the economic structure: An input-output approach. *The Review of Economics and Statistics*, 262–271.
54. Lim, S., Quercia, D., & Finkelstein, A. (2010). StakeNet: using social networks to analyse the stakeholders of large-scale software projects. In *Proceedings of the 32nd ACM/IEEE International Conference on Software Engineering* (Vol. 1, pp. 295–304).
55. Malczewski, J., & Rinner, C. (2015). *Multicriteria decision analysis in geographic information science*. Springer.
56. McGillivray, M. (1991). The human development index: Yet another redundant composite development indicator? *World Development, 19*(10), 1461–1468.
57. Moss, S., Pahl-Wostl, C., & Downing, T. (2001). Agent-based integrated assessment modelling: The example of climate change. *Integrated Assessment, 2*(1), 17–30.
58. Ness, B., Urbel-Piirsalu, E., Anderberg, S., & Olsson, L. (2007). Categorising tools for sustainability assessment. *Ecological Economics, 60*(3), 498–508.
59. Office for National Statistics. (2013, July). Sustainable development indicators [Computer software manual]. London, UK: Author.
60. OpenLCA. (2016). http://www.openlca.org/.
61. Pezzey, J. (1989). *Definitions of sustainability*. UK Centre for Economic and Environmental Development.
62. Pizzol, M., Scotti, M., & Thomsen, M. (2013). Network analysis as a tool for assessing environmental sustainability: Applying the ecosystem perspective to a Danish Water Management System. *Journal of Environmental Management, 118*, 21–31.
63. Pope, J., Annandale, D., & Morrison-Saunders, A. (2004). Conceptualising sustainability assessment. *Environmental Impact Assessment Review, 24*(6), 595–616.
64. Prell, C., Hubacek, K., & Reed, M. (2009). Stakeholder analysis and social network analysis in natural resource management. *Society and Natural Resources, 22*(6), 501–518.
65. Rotmans, J., & Asselt, M. (1996). Integrated assessment: A growing child on its way to maturity. *Climatic Change, 34*(3–4), 327–336.
66. Schaltegger, S., Bennett, M., & Burritt, R. (2006). *Sustainability accounting and reporting* (Vol. 21). Springer Science & Business Media.
67. SETAC. (2016). *Society of environmental toxicology and chemistry*. http://www2.setac.org/.

68. Shannon, C. (2001). A mathematical theory of communication. *ACM SIGMOBILE Mobile Computing and Communications Review*, *5*(1), 3–55.
69. Silalertruksa, T., & Gheewala, S. H. (2009). Environmental sustainability assessment of bioethanol production in Thailand. *Energy*, *34*(11), 1933–1946.
70. Simon, H. (1991). The architecture of complexity. In *Facets of systems science* (pp. 457–476). Springer.
71. Smith, C., & McDonald, G. (1997). Assessing the sustainability of agriculture at the planning stage. *Journal of Environmental Management*, *52*(1), 15–37.
72. Sparrevik, M., Barton, D., Bates, M., & Linkov, I. (2012). Use of stochastic multicriteria decision analysis to support sustainable management of contaminated sediments. *Environmental Science & Technology*, *46*(3), 1326–1334.
73. Statistical Office of the European Communities. (1982). *Eurostatistik, Daten zur Konjunkturanalyse/[Statistisches Amt der Europaischen Gemeinschaften] = Eurostatistics, data for short-term economic analysis / [Statistical Office of the European Communities]*. Office for Official Publications of the European Communities Luxembourg.
74. Sterman, J., & Rahmandad, H. (2014). *Introduction to system dynamics*. Massachusetts Institute of Technology: MIT OpenCourseWare. Retrieved from http://ocw.mit.edu. License: Creative Commons BY-NC-SA.
75. Tabara, J., Roca, E., Madrid, C., Valkering, P., Wallman, P., & Weaver, P. (2008). Integrated sustainability assessment of water systems: Lessons from the Ebro River Basin. *International Journal of Innovation and Sustainable Development*, *3*(1–2), 48–69.
76. Tibor, T., & Feldman, I. (1996). *ISO 14000: A guide to the new environmental management standards*. Chicago: IL (USA) Irwin.
77. Todorov, V., & Marinova, D. (2011). Modelling sustainability. *Mathematics and Computers in Simulation*, *81*(7), 1397–1408.
78. Varga, L., Allen, P., Strathern, M., Rose-Anderssen, C., Baldwin, J., & Ridgway, K. (2009). Sustainable supply networks: A complex systems perspective. *Emergence: Complexity and Organization*, *11*(3), 16.
79. Videira, N., Antunes, P., Santos, R., & Lopes, R. (2010). A participatory modelling approach to support integrated sustainability assessment processes. *Systems Research and Behavioral Science*, *27*(4), 446–460.
80. Wang, J., Jing, Y., Zhang, C., & Zhao, J. (2009). Review on multi-criteria decision analysis aid in sustainable energy decision-making. *Renewable and Sustainable Energy Reviews*, *13*(9), 2263–2278.
81. Weisz, H., & Duchin, F. (2006). Physical and monetary input–output analysis: What makes the difference? *Ecological Economics*, *57*(3), 534–541.
82. Whitmarsh, L., & Nykvist, B. (2008). Integrated sustainability assessment of mobility transitions: Simulating stakeholders' visions of and pathways to sustainable land-based mobility. *International Journal of Innovation and Sustainable Development*, *3*(1–2), 115–127.
83. Xu, Z. (2011, July). Application of System Dynamics model and GIS in sustainability assessment of urban residential development. In *Proceedings of the 29th International Conference of the System Dynamics Society*. Washington, DC.
84. Yang, X. (2010). *Applying stochastic programming models in financial risk management*. The University of Edinburgh.
85. Zadeh, L. (1965). Fuzzy sets. *Information and Control*, *8*(3), 338–353.
86. Zitrou, A., Bedford, T., & Daneshkhah, A. (2013). Robustness of maintenance decisions: Uncertainty modelling and value of information. *Reliability Engineering & System Safety*, *120*, 60–71.
87. Zou, Z., Yun, Y., & Sun, J. (2006). Entropy method for determination of weight of evaluating indicators in fuzzy synthetic evaluation for water quality assessment. *Journal of Environmental Sciences*, *18*(5), 1020–1023.

Sustainable Maintenance Strategy Under Uncertainty in the Lifetime Distribution of Deteriorating Assets

Alireza Daneshkhah, Amin Hosseinian-Far and Omid Chatrabgoun

Abstract In the life-cycle management of systems under continuous deterioration, studying the sensitivity analysis of the optimised preventive maintenance decisions with respect to the changes in the model parameters is of a great importance. Since the calculations of the mean cost rates considered in the preventive maintenance policies are not sufficiently robust, the corresponding maintenance model can generate outcomes that are not robust and this would subsequently require interventions that are costly. This chapter presents a computationally efficient decision-theoretic sensitivity analysis for a maintenance optimisation problem for systems/structures/assets subject to measurable deterioration using the Partial Expected Value of Perfect Information (PEVPI) concept. Furthermore, this sensitivity analysis approach provides a framework to quantify the benefits of the proposed maintenance/replacement strategies or inspection schedules in terms of their expected costs and in light of accumulated information about the model parameters and aspects of the system, such as the ageing process. In this paper, we consider random variable model and stochastic Gamma process model as two well-known probabilistic models to present the uncertainty associated with the asset deterioration. We illustrate the use of PEVPI to perform sensitivity analysis on a maintenance optimisation problem by using two standard preventive maintenance policies, namely age-based and condition-based maintenance policies. The optimal strategy of the former policy is the time of replacement or repair and the optimal strategies of the later policy are the inspection time and the preventive maintenance ratio. These optimal strategies are determined by minimising the corresponding expected cost rates for the given deterioration

A. Daneshkhah (✉)
The Warwick Centre for Predictive Modelling, School of Engineering,
The University of Warwick Coventry, Coventry CV4 7AL, UK
e-mail: ardaneshkhah@gmail.com; a.daneshkhah@warwick.ac.uk

A. Hosseinian-Far
School of Computing, Creative Technologies & Engineering,
Leeds Beckett University, Leeds LS6 3QR, UK
e-mail: A.Hosseinian-Far@leedsbeckett.ac.uk

O. Chatrabgoun
Department of Statistics, Faculty of Mathematical Sciences and Statistics,
Malayer University, Malayer, Iran
e-mail: O.Chatrabgoun@malayeru.ac.ir

© Springer International Publishing AG 2017
A. Hosseinian-Far et al. (eds.), *Strategic Engineering for Cloud Computing and Big Data Analytics*, DOI 10.1007/978-3-319-52491-7_2

models' parameters, total cost and replacement or repair cost. The robust optimised strategies to the changes of the models' parameters can be determined by evaluating PEVPI's which involves the computation of multi-dimensional integrals and is often computationally demanding, and conventional numerical integration or Monte Carlo simulation techniques would not be helpful. To overcome this computational difficulty, we approximate the PEVPI using Gaussian process emulators.

Keywords Deterioration models · Partial Expected Value of Perfect Information · Gaussian process · Optimised maintenance · Cost-benefit

1   Introduction

The resilience of an asset/machinery component or networked infrastructure system is greatly dependent on an efficient cost effective life-cycle management. This can be achieved by determining optimal maintenance and rehabilitation scheduling schemes. Maintenance costs for an asset or a networked infrastructure systems including rail, water, energy, bridge etc. are swiftly rising, but current estimates suggest that ontime optimised maintenance schedules could save one trillion dollars per year on infrastructure costs [10].

The maintenance strategies have generally been divided into two categories: Corrective Maintenance (CM); and Preventative Maintenance (PM). The former includes repairing failed components and systems, while the latter involves systematic inspection and correction of initiative failures, before they progress into major faults or defects. In the recent years, an increasing dominance of PM has been clearly observed with overall costs illustrated to be lower than CM strategy. The preventive maintenance is extensively applied to lessen asset deterioration and mitigate the risk of unforeseen failure. This maintenance strategy can be further classified into two methods: Time-Based Maintenance (TBM), and Condition-Based Maintenance (CBM). In the TBM, maintenance activities take place at predetermined time intervals, but in the CBM, interventions are immediately carried out based on the information collected through condition sensing and monitoring processes (either manual or automated). Both TBM and CBM are widely used for asset/infrastructure life-cycle management decision making, and extensively studied in [1, 9, 17, 28].

The main difficulty to make informed PM decisions is that predicting the time to first inspection, maintenance intervention, or replacement is confounded by model parameters' uncertainties associated with the failure, deterioration, repair, or maintenance distributions. As a result, studying sensitivity of the model output with respect to the changes in the model parameters/inputs is very essential to determine an optimal maintenance strategy under these uncertainties. One of the main aims of this paper is to investigate sensitivity analysis of the optimised maintenance with respect to the changes in the model's inputs when the aforementioned preventive maintenance strategies are considered for the asset/infrastructure which is under continuous deterioration. The optimal maintenance decision under this maintenance policies is normally considered as the inspection interval and the preventive maintenance ratio that would minimize the expected total cost of the maintenance strategy. This type of maintenance strategy is known to practically be more useful, particularly for larger and more complex systems [1, 29]. This is mainly because it eliminates the need to record component ages. It should be noted that finding the optimal decision under a CBM policy for a deteriorating component involves solving a two-dimensional optimisation problem, while for the TBM case the aim is to determine the critical age as a single strategy variable.

As mentioned above, the PM policy cost function is dominated by the deterioration and repair distribution's parameters. Therefore, the computation of a mean cost rates for a specific PM policy is not sufficiently robust, and the corresponding maintenance model can generate results that are not robust. In other words, the determination of an optimal maintenance intervention will be sensitive to the parameters creating uncertainty as to the optimal strategy. The uncertainty around the optimal PM maintenance can be mitigated by gathering further information on some (or all) of the model parameters/inputs. In particular, Partial Expected Value of Perfect Information (PEVPI) computations provide an upper bound for the value (in terms of cost-benefit) that can be expected to be yielded from removing uncertainty in a subset of the parameters to the cost computation. The PEVPI provides a decision-informed sensitivity analysis framework which enables researchers to determine the key parameters of the problem and quantify the value of learning about certain aspects of the system [21, 29]. In maintenance studies [7, 11], this information can play a crucial role, where we are interested in not only determining an optimal PM strategy, but also in collecting further information about the system features, including the deterioration process to make more robust decisions.

The identification of PEVPI requires the computation of high-dimensional integrals that are regularly expensive to evaluate, and typical numerical integration or Monte Carlo simulation techniques are not practical. This computational burden can be overcome, by applying the sensitivity analysis through the use of Gaussian process (GP) emulators [21, 29]. In this paper, we adopt and execute this sensitivity analysis approach for determining robust optimised PM strategies for a component/system which is under continuous deterioration.

We use the random variable model (RV) to probabilistically model the deterioration of a component/system of interest. Under the RV model [22], the probability density and distribution functions of the lifetime are respectively given by

$$f_T^R(t) = \frac{(\delta/\rho)}{\Gamma(\eta)}(\frac{\rho}{\delta t})^{\eta+1}e^{-\rho/(\delta t)}, \tag{1}$$

and

$$F_T^R(t) = 1 - \mathscr{G}(\rho/t; \eta, \delta) = 1 - \mathscr{G}(\rho; \eta, \delta t), \tag{2}$$

where η and δ are, respectively, the shape and scale parameters, $\mathscr{G}(\rho/t; \eta, \delta)$ denote the gamma cumulative distribution function with the same shape and scale parameters for $X = \rho/T$, and $\rho = (r_0 - s) > 0$ is called the available design margin or a *failure threshold.* In the last expression, r_0 is the initial resistance of the component against the load effect, s. Thus, a failure is defined as the event at which the cumulative amount of deterioration exceeds the deterioration threshold (ρ). The threshold ρ, s and r_0 are assumed to be deterministic constants for simplicity of discussion (see [22] for further details).

The rest of the chapter is organised as follows. In Sect. 2, we discuss how this probabilistic deterioration model links to TBM and CBM maintenance optimisation problems. We formulate uncertainty quantification of the optimised PM policies using the decision-informed sensitivity analysis in Sect. 3. The GP emulator required to compute PEVPI's as within the context of decision-theoretic sensitivity analysis is also briefly discussed in Sect. 3. Section 4 is dedicated to derive the robust optimised maintenance decisions for TBM and CBM policies using several illustrative settings of different complexity. We conclude by discussing the implications of our approach and identify opportunities for future work.

2 Optimal Preventive Maintenance Policy

The central objective of a preventive maintenance (TBM or CBM) optimisation model is to determine the value of the decision variable T (replacement time or inspection time) that optimises a given objective function amongst the available alternative maintenance decisions. For instance in a TBM policy, the optimisation problem is usually defined over a finite time horizon $[0, t]$, and the objective function, denoted by $C(t)$, is the long-term average cost. There are various ways to define these costs [3, 8]. The long-term mean cost per unit of time is normally defined in terms of the length of two consecutive replacements (or life cycle) as follows:

$$\mathscr{C}(T) = \frac{C(T)}{L(T)}. \tag{3}$$

The following formula is an example of the expected cost per unit of a component under a general TBM policy

$$\mathscr{C}(T) = \frac{c_1 F(T) + c_2 R(T)}{T \cdot R(T) + \int_0^T tf(t)\mathrm{d}t + \tau}, \tag{4}$$

where $F(T)$ is the failure distribution function of a system at time T (or probability of unplanned replacement due to an unexpected failure), $R(T) = 1 - F(T)$ is the probability of planned replacement at time T, c_1 is the cost of a corrective maintenance, c_2 is the cost of planned replacement and τ is the expected duration of replacement.

The objective is then to identify the optimal strategy T^* that corresponds to the minimum cost rate (cost per unit of time), that is

$$T^* = \arg\min_{T>0}\{\mathscr{C}(T)\}. \tag{5}$$

A similar methods is used to determine the optimised CBM strategy. The cost function in this policy is the mean cost rate which is defined as

$$\mathscr{K}(t_I, v) = \frac{E[C(t_I, v)]}{E[L(t_I, v)]}, \tag{6}$$

where $E[C(t_I, v)]$ is the renewal cycle cost, $E[L(t_I, v)]$ is the renewal cycle length, t_I is the inspection time interval and v is the PM ratio. The details of numerator and denominator of the mean cost rate will be given in Sect. 4.

The objective is then to find t_I^* and v^* so that $\mathscr{K}(t_I^*, v^*)$ becomes the minimal cost solution.

2.1 Uncertainty Quantification Using Via Decision-Theoretic Sensitivity Analysis

The optimal maintenance strategies derived by minimising the expected cost rate is influenced by characteristics such as the deterioration process or failure behaviour of the system and the characteristics of maintenance tasks (including repair/replacement policy, maintenance crew and spare part availability etc.). These characteristics are subject to uncertainty, prompting study of the sensitivity of an optimal maintenance strategy with respect to changes in the model parameters and other inflecting uncertain inputs. Such an analysis improves understanding of the 'robustness' of the derived inferences or predictions of the model, and, offers a tool for determining the critical influences on model predictions [6, 27]. Zitrou et al. [29] summarise the main sensitivity measures and discuss their values and applications in an extensive sensitivity analysis. They conclude that a simple yet effective method of implementing sensitivity analysis is to vary one or more parameter inputs over some plausible range, whilst keeping the other parameters fixed, and then examine the effects of these changes on the model output. Although this method is straightforward to implement and interpret, it becomes inconvenient where there are large numbers of model parameters or when the model is computationally intensive.

In order to resolve this difficulty, we use a variance-based method for sensitivity analysis [27]. This approach can capture the fractions of the model output variance which are explained by the model inputs. In addition, it can also provide the total contribution to the output variance of a given input—i.e. its marginal contribution and cooperative contribution. The contribution of each model's input to the model output variance serves as an indicator of how strong an influence a certain input or parameter has on model output variability. However, within a decision-making context like the maintenance optimisation problem, we are primarily interested in the effect of parameter uncertainty on corresponding utility or loss. To achieve this objective, we use the concept of the Expected Value of Perfect Information (EVPI) as a measure of parameter importance [21, 29]. The EVPI approach allows the application of sensitivity analysis to the maintenance optimisation model and identifies the model parameters for which collecting additional information (learning) prior to the maintenance decision would have a significant impact on total cost.

3 Decision-Theoretic Sensitivity Analysis

3.1 Expected Value of Perfect Information

As discussed in the previous sections, the objective function of interest to us is the expected cost function (e.g. the cost rate function given in Equation (4) for TBM or the mean cost rate given in (6) for CBM). These cost functions take reliability and maintenance parameters as uncertain inputs (denoted by θ) and a decision parameter, T (which could be critical age or periodic inspection interval). A strategy parameter (which is fixed) needs to be selected in the presence of unknown reliability and maintenance parameters. These unknown parameters can be modelled by a joint density function, $\pi(\theta)$. In the maintenance optimisation setting, the decision-maker can choose the strategy parameter T (from a range or set of positive numbers) where each value of T corresponds to a maintenance decision. The decision T is selected so that the following utility function is maximised

$$U(T, \theta) = -\mathscr{C}(T; \theta), \tag{7}$$

where $\mathscr{C}(T; \theta)$ is a generic cost function per unit of time given the unknown parameters θ.

Suppose that we need to make a decision now, on the basis of the information in $\pi(\theta)$ only. The optimal maintenance decision (known as *baseline* decision), given no additional information, has expected utility

$$U_0 = \arg \max_{T>0} E_\theta [U(T, \theta)], \tag{8}$$

where

$$E_\theta [U(T, \theta)] = - \int_\theta \mathscr{C}(T; \theta)\pi(\theta)\mathrm{d}\theta. \tag{9}$$

Now suppose that we wish to learn the precise value of a parameter θ_i in θ before making a decision (e.g. through exhaustive testing; new evidence elicited from the domain expert). Given θ_i, we are still uncertain about the remaining input parameters, $\underline{\theta_i} = (\theta_1, \dots, \theta_{i-1}, \theta_{i+1}, \dots, \theta_n)$, and so we would choose the maintenance strategy to maximise

$$E_{\theta|\theta_i}[U(T,\theta)] = -\int_{\theta_i} \mathscr{C}(T;\theta)\pi(\theta \mid \theta_i)d\theta_i. \tag{10}$$

The expected utility of learning θ_i is then given by:

$$U_{\theta_i} = E_{\theta_i}\left[\arg\max_{T>0} E_{\theta|\theta_i}\{U(T,\theta)\}\right]. \tag{11}$$

Therefore, learning about parameter θ_i before any maintenance decision being taken will benefit the decision-maker by:

$$\text{EVPI}_{\theta_i} = E_{\theta_i}[U_{\theta_i}] - U_0. \tag{12}$$

Therefore, the quantity EVPI_{θ_i}, known as the partial Expected Value of Perfect Information (partial EVPI or PEVPI), is a measure of the importance of parameter θ_i in terms of the cost savings that further learning (data collection) will achieve.

EVPI's allow for sensitivity analysis to be performed in a decision-theoretic context. However, the computation of partial EVPIs as in (12) requires the evaluation of expectations of utilities over many dimensions. Whereas the one-dimensional integral $E_{\theta_i}[U_{\theta_i}]$ can be evaluated efficiently using Simpson's rule, averaging over the values of multiple parameters is computationally intensive. One way to approximate these expectations is to use a Monte Carlo numerical method. However, the Monte Carlo-based integration methods require a large number of simulations which make the computation of the PEVPI's impractical. Zitrou et al. [29] propose an alternative method for resolving this problem by utilising a GP emulator-based sensitivity analysis to the objective function of interest. This method enables computation of the multi-dimensional expectations at a limited number of model evaluations with relative computational ease. We develop this method further for the purposes mentioned above.

3.2 Gaussian Process Emulators

An emulator is an approximation of a computationally demanding model, referred to as the *code*. An emulator is typically used in place of the code, to speed up calculations. Let $\mathscr{C}(\cdot)$ be a code that takes as input a vector of parameters $\theta \in \mathscr{Q} \subset \mathbb{R}^q$, for some $q \in \mathbb{Z}_+$, and has output $y = \mathscr{C}(\theta)$, where $y \in \mathbb{R}$. As we will see later on, this is not a restrictive assumption, and we will let $y \in \mathbb{R}^s$, for some $s \in \mathbb{Z}_+$. For the time being, let $\mathscr{C}(\cdot)$ be a deterministic code, that is for fixed inputs, the code produces the same output each time it 'runs'.

An emulator is constructed on the basis of a sample of code runs, called the *training set*. In a GP emulation context, we regard $\mathscr{C}(\cdot)$ as an unknown function, and use a $q-$ dimensional GP to represent prior knowledge on $\mathscr{C}(\cdot)$, i.e.

$$\mathscr{C}(\cdot) \sim N_q(m(\cdot), v(\cdot, \cdot)). \tag{13}$$

We subsequently update our knowledge about $\mathscr{C}(\cdot)$ in the light of the training set, to arrive at a posterior distribution of the same form.

Expression (13) implies that for every $\{\theta_1, \ldots, \theta_n\}$ output $\{\mathscr{C}(\theta_1), \ldots \mathscr{C}(\theta_n)\}$ has a prior multivariate normal distribution with mean function $m(\cdot)$ and covariance function $v(\cdot, \cdot)$. There are many alternative models for the mean and covariate functions $m(\cdot)$. Here, we use the formulation in line with [19], and assume

$$m(\theta) = h(\theta)^\mathsf{T} \beta \tag{14}$$

for the mean function, and

$$v(\theta, \theta') = \sigma^2 c(\theta, \theta') \tag{15}$$

for the covariance function. In 14, $h(\cdot)$ is a vector of q known regression functions of θ and β is a vector of coefficients. In (15), $c(\cdot, \cdot)$ is a monotone correlation function on $\mathbb{R}^+$ with $c(\theta, \theta) = 1$ that decreases as $|\theta - \theta'|$ increases. Furthermore, the function $c(\cdot, \cdot)$ must ensure that the covariance matrix of any set of outputs is positive semi-definite. Throughout this paper, we use the following correlation function which satisfies the aforementioned conditions and is widely used in the Bayesian Analysis of Computer Code Outputs (BACCO) emulator ([18, 21]) for its computational convenience:

$$c(\theta, \theta') = \exp\{-(\theta - \theta')^\mathsf{T} B(\theta - \theta')\}, \tag{16}$$

where B is a diagonal matrix of positive smoothness parameters. B determines how close two inputs θ and θ' need to be such that the correlation between $k(\theta)$ and $\mathscr{C}(\theta')$ takes a particular value. For further discussion on such modelling issues [14]. To estimate parameters β and σ, we use a Bayesian approach as in [20]: a normal inverse gamma prior for (β, σ^2) is updated in the light of an observation vector $y = (\mathscr{C}(\theta_1), \ldots, \mathscr{C}(\theta_n))^\mathsf{T}$, to give a GP posterior (see [29] for the details of how the posterior distribution of the function of interest can be derived). Note that y is obtained by running the initial code $\mathscr{C}(\cdot)$ n times on a set of design points $(\theta_1, \theta_2, \ldots, \theta_n)^\mathsf{T}$.

In many cases, input parameters are subject to uncertainty and are modelled as random variables. The input of the code is now a random vector θ with probability distribution F, implying that the code output $Y = \mathscr{C}(\theta)$ is also a random variable. The uncertainty in the output is epistemic, arising from the uncertainty of the input parameters. But there is also uncertainty due to the incomplete knowledge of the model output, called *code uncertainty* (we are not running the actual code, but just an approximation). We can quantify code uncertainty on the basis of the covariance function (15) and control its effect by modifying the number of design points.

Emulators are useful tools for uncertainty and sensitivity analysis [15, 18]. For GP emulators in particular, this is due to the fact that Bayesian quadrature as described in

[19] allows one to take advantage of the emulator's analytical properties to evaluate the expected value $E[Y]$ and the variance $Var[Y]$ relatively quickly. In particular, since Y is a GP, the integral

$$E[Y] = \int_\theta \mathscr{C}(\theta)\pi(\theta)d\theta \tag{17}$$

has a normal distribution. In this particular context, we are interested in computing partial EVPIs as in (12). By using an emulator to approximate utility $U(T, \theta)$, expectations of the utility can be computed rapidly, considerably reducing the computational burden of sensitivity analysis. Emulators perform better than standard Monte Carlo methods in terms of both accuracy of model output and computational effort [14, 18].

The objective of the maintenance optimisation problem described here is to identify decision T (time where PM is performed) that maximises the Utility given in (7). Essentially, there are two approaches to the problem:

A assume that T belongs to a finite set $\mathscr{S}$ comprising s options $T_1, T_2, \ldots, T_s$, or
B assume that T can take any value in $(T_{\min}, T_{\max})$.

Under Approach A, to identify the optimal decision we need to determine the vector of outputs $\mathbf{Y} = (Y_1, \ldots, Y_s)$, where $Y_j = U(T_j, \theta)$ $(j = 1, \ldots s)$ is described in (7). As discussed in [5], there are essentially two procedures for emulating a code with a multi-dimensional output like this one: the *Multi-Output emulator (MO)* and the *Many Single Outputs (MS)* emulator. Alternatively, under Approach B, we need to determine output $Y(T) = U(T, \theta)$ as a function of decision variable T. To do so, one can use a *Time Input (TI)* emulator [5].

The MO emulator is a multivariate version of the single-output emulator, where the dimension of the output space is s. This process allows for the representation of any correlations existing among the multiple outputs. The MS emulator procedure treats $\{Y_1, \ldots, Y_s\}$ as independent random variables, and emulates each output Y_j separately. This means that s separate GP emulators are built, each describing the utility for each decision $T \in \mathscr{S}$. Finally, the TI emulator is a single-output emulator that considers decision variable T as an additional input parameter. The advantage of this approach is that S does not have to be a finite space, and utility $U(T, \theta)$ can be determined for any value of T over some interval $(T_{\min}, T_{\max})$. For the maintenance optimisation problem examined here, the TI emulator has a very important advantage over the other two: it allows for decision T to be a continuous variable. Expectations $E[Y]$ are continuous functions of T, and the utilities of the optimal strategies are calculated without restricting the decision-maker to choose amongst a predetermined, finite number of options. We believe that this feature outweighs the more general correlation structure provided by the MO emulator. We note, however, that this may not be the case in dynamic codes, where the representation of the temporal structure of some physical process is key.

3.3 The TI Emulator

Suppose that the optimal decision T in a maintenance optimisation problem (critical age or periodic interval) belongs to an infinite set $S = (T_{min}, T_{max})$. We consider T as a code input and we are interested in building a single-output emulator to approximate code

$$k(T, \boldsymbol{\theta}) = U(T, \boldsymbol{\theta}) = -\mathscr{C}(T; \theta), \tag{18}$$

where $\mathscr{C}(T; \theta)$ is the cost rate function. This will allow us to calculate expected utilities $E_{\boldsymbol{\theta}}[U(T, \boldsymbol{\theta})]$ and $E_{\boldsymbol{\theta}|\theta_i}[U(t, \boldsymbol{\theta})]$ for $T \in \mathscr{S}$—see Relationships (8) and (10)—fast and efficiently. We note that this setting is an extension of the decision problem considered in [21], where the decision-maker must choose among a small set of options, i.e. where the optimal decision T belongs to a finite set S.

To estimate the hyper-parameters of the TI emulator, we generate training set $\mathscr{T}$ consisting of code outputs

$$y_1 = k(\boldsymbol{x_1}), \dots, y_N = k(\boldsymbol{x_N}),$$

where $(\boldsymbol{x_1}, \boldsymbol{x_2}, \dots, \boldsymbol{x_N})^\mathsf{T}$ are design points. We have

$$x_l = (T_i, \boldsymbol{\theta_j}), \quad l = 1, 2, \dots, N = s \times n,$$

where T_i is a maintenance decision ($i = 1, \dots, s$) and $\boldsymbol{\theta_j}$ are (reliability, maintainability) parameter values ($j = 1, \dots, n$).

The choice of design points affects how well the emulator is estimated. Here, we choose equally spaced points $\{T_1, \dots, T_s\}$ so that interval S is properly covered. Points $(\boldsymbol{\theta_1}, \boldsymbol{\theta_2}, \dots, \boldsymbol{\theta_n})^\mathsf{T}$ are generated using Latin hypercube sampling (see [16]), which ensures that the multi-dimensional parameter space is sufficiently covered.

As mentioned earlier, building a TI emulator requires the inversion of an $N \times N$ matrix. Given the size of the training set, this can be computationally challenging. Essentially, there are two ways to build the TI emulator: (1) fit a GP directly to the whole training set $\mathscr{T}$ obtained as described above; (2) separate $\mathscr{T}$ and fit two GPs: one on the set of design points $(\boldsymbol{\theta_1}, \boldsymbol{\theta_2}, \dots, \boldsymbol{\theta_n})$ and one on the time input data points $\{T_1, \dots, T_s\}$ [5, 25].

Zitrou et al. [29] present the methodology to fit these two GPs to a similar decision-informed optimization problem, including estimating the roughness (or smoothness) parameters and other hyper-parameters. They showed that however the firs approach based on fitting a single GP to the whole training set $\mathscr{T}$ takes longer, but it would produce more accurate results. They showed that the relative mean squared error of the posterior predictive mean of the first model (based on fitting a single GP) is much smaller than when fitting two GP. We therefore follow their suggestion and fit a single GP to the full training set.

The baseline maintenance strategy is then chosen as a value of T that maximises expected utility as

$$U_0 = \max_{T \in \mathcal{S}} E_k \{ E_\theta [k(T, \boldsymbol{\theta})] \} \tag{19}$$

and the utility of the optimal strategy in (10), after learning the value of θ_i, becomes

$$U_{\Theta_i} = \max_{T \in \mathcal{S}} E_k \{ E_{T, \theta | \Theta_i} [k(T, \boldsymbol{\theta})] \}. \tag{20}$$

Bayesian quadrature [19] allows us to compute these expectations relatively quickly based on the fitted GP and calculate PEVPIs for the different parameters as in (12). The details of the approximation of this type of integral (expectation) in terms of the fitted GP can be found in [6].

We use R and GEM-SA packages to fit the GP to the training points and then approximate the expected utilities and their corresponding uncertainty bounds. To calculate the aforementioned expected utilities, the calculations are carried out based on the discretisation of the interval $\mathcal{S}$ (maintenance decision) and the support of the joint prior distribution of the parameters $\pi(\boldsymbol{\theta})$. It is apparent that the computation of these expectation can become quite expensive by choosing a finer discretisation. The following section presents two illustrative examples. The focus here is on the way emulators can be used to perform sensitivity analysis based on EVPI, providing a resource efficient method for maintenance strategy identification and identifying targets for institutional learning (uncertainty reduction). In the first example, we build an emulator for a TBM optimisation problem and in the second example finding a robust CBM of a civil structure using emulator-based sensitivity analysis is presented.

4 Numerical Examples

4.1 Sensitivity Analysis of Time-Based Maintenance Decisions

Under the TBM policy (also known as age-based replacement), the system or component under study is in one out of two operating conditions; working or failed. System failure is identified immediately and corrective maintenance (CM) actions are undertaken to restore the system to its original condition. Regardless of the system condition, the system is renewed when it reaches a predetermined time (or age) T^*. In the TBM optimisation problem, the main challenge is to identify the optimal time to maintenance to minimise overall maintenance costs. This optimisation problem is usually defined over a finite horizon $[0, t]$, and we seek to minimise the objective cost function $\mathcal{C}(t)$ over this time interval.

Using the renewal theory [3, 8], the expected cost rate per unit of time can be mathematically expressed as

$$\mathscr{C}_R(t;\theta_1) = \frac{C_F F_T(t;\theta_1) + C_P[1 - F_T(t;\theta_1)]}{\int_0^t [1 - F_T(t;\theta_1)]dt}, \tag{21}$$

where C_F is the total cost associated with all the consequences of system failure (particularly, structural failure due to deterioration), C_P is the cost associated with a preventive maintenance action, $F_T(t;\theta_1)$ is the system lifetime distribution function given in (2), and θ_1 is the set of the lifetime distribution parameters.

The cost rate function given in (21) can be developed further (see [2, 3]) to

$$\mathscr{C}_R(t;\theta) = \frac{C_1 F_T(t;\theta_1) + C_2[1 - F_T(t;\theta_1)]}{\int_0^t [1 - F_T(t)]dt + \tau_r(\theta_2)}, \tag{22}$$

where $\tau_r(\theta_2)$ is the expected duration of the maintenance action, and is defined by

$$\tau_r(\theta_2) = \int_0^\infty t g_T(t;\theta_2)dt, \tag{23}$$

where $g_T(t;\theta_2)$ is the time to repair (or replacement) distribution, θ_2 is the set of repair distribution parameters, and $\theta = (\theta_1,\theta_2)$. Without loss of generality, $g_T(t;\theta_2)$ is assumed to follow a Gamma distribution with α and β as shape and scale parameters, respectively.

For numerical illustration, we set $C_F = 50$ and $C_p = 10$. As discussed in Sect. 1, the lifetime distribution of a deteriorating component using the RV model follows an inverted Gamma distribution with the density function given in (1) In this distribution, η and δ are, respectively, the shape and scale parameters of the lifetime distribution. Figure 1 illustrates how the expected cost rates change over the decision variable T for specific values of parameters, $\theta = (\eta, \delta, \alpha, \beta)$.

Fig. 1 Total long-run average costs per unit time function for different values of $\theta = (\eta, \delta, \alpha, \beta)$

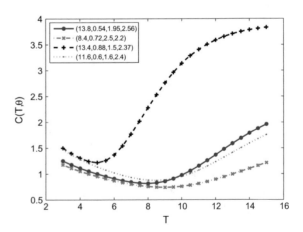

The decision-maker proposes the following prior distribution on θ

$$\pi(\theta) = \pi_1(\eta)\pi_2(\delta)\pi_3(\alpha)\pi_4(\beta), \tag{24}$$

where each of these parameters individually is uniformly distributed as follows:

$$\eta \sim U(3, 14), \quad \delta \sim U(0.15, 0.95), \quad \alpha \sim U(1, 3), \quad \beta \sim U(2, 3),$$

where $U(a_1, b_1)$ denote a uniform density function defined over (a_1, b_1).

It can be shown that the cost function in (22) has a unique optimal solution (according to Theorem 1 given in [2]). When the uncertainty in input parameters θ are included, the optimal maintenance decision will lie in the interval, $I = [3, 5]$ [29].

In order to lower the computational load of computing the value of information measures (PEVPIs) as the sensitivity analysis index, a TI emulator is fitted to the cost rate function $\mathscr{C}_R(t; \theta)$, given in (22). The total training data-points to build this emulator is 1500 selected as follows. We first generate 60 design points from the joint prior distribution, θ, using the Latin hypercube design [26]. We then calculate the cost rate function (as a computer code) at each design point for 25 values of T (i.e. $T = 3, 3.5, 4, \dots, 15$).

Using the fitted GP, the baseline optimal decision is derived at $T = 8.5$ where the corresponding maximum utility is $U_0 = 0.896$. So, if there is no additional information available on individual input parameters, apart from the prior information, the optimal time to maintenance is at 8.5 time units resulting in an almost 90% cost saving compare the corrective maintenance. Further saving can be achieved if additional information about the values of the parameters can be provided before making any decision. For example, suppose that α is known before making a decision. Table 1 provides the detailed information about the optimal decisions for the different values of η. For instance, when the shape parameter of the lifetime distribution of a component under study takes values in $(3, 3.825)$, then the cost rate is minimum for $T = 10.2$, but if $\eta \in (8.255, 8.775)$, then the optimal maintenance decision

Table 1 Optimal decisions when parameter η is known prior the maintenance decision

Range	T	Range	T
(3, 3.825)	10.2	(8.225, 8.775)	7.80
(3.825, 4.375)	11.4	(8.775, 9.875)	7.2
(4.375, 4.925)	13.8	(9.875, 10.975)	7.80
(4.925, 5.475)	10.80	(10.975, 11.525)	6.6
(5.475, 6.025)	7.20	(11.525, 12.075)	7.2
(6.025, 7.125)	6.6	(12.075, 13.175)	13.2
(7.125, 7.675)	7.80	(13.175, 13.725)	10.8
(7.675, 8.225)	8.40	(13.725, 14)	9

Table 2 Optimal decisions when parameter δ is known prior the maintenance decision

Range	T	Range	T
(0.15, 0.29)	7.2	(0.61, 0.73)	9
(0.29, 0.43)	7.8	(0.73, 0.93)	9.6
(0.43, 0.61)	8.4	(0.93, 0.95)	10.2

Table 3 Optimal decisions when parameter α is known prior the maintenance decision

Range	T	Range	T
(1, 1.10)	8.52	(2.30, 2.34)	8.28
(1.10, 1.34)	8.28	(2.34, 2.38)	8.52
(1.34, 1.50)	8.04	(2.38, 2.42)	9
(1.50, 1.66)	7.8	(2.42, 2.46)	9.48
(1.66, 1.82)	7.56	(2.46, 2.50)	9.96
(1.82, 2.14)	7.32	(2.50, 2.54)	10.20
(2.14, 2.22)	7.56	(2.54, 2.62)	10.44
(2.22, 2.26)	7.80	(2.62, 2.78)	10.68
(2.26, 2.30)	8.04	(2.78, 3)	10.92

Table 4 Optimal decisions when parameter β is known prior the maintenance decision

Range	T	Range	T
(2, 2.03)	14.76	(2.19, 2.21)	9.48
(2.03, 2.05)	13.80	(2.21, 2.23)	9.24
(2.05, 2.07)	13.32	(2.23, 2.27)	9
(2.07, 2.09)	12.60	(2.27, 2.33)	8.76
(2.09, 2.11)	11.64	(2.33, 2.41)	8.52
(2.11, 2.13)	10.68	(2.41, 2.55)	8.28
(2.13, 2.15)	10.20	(2.55, 2.77)	8.04
(2.15, 2.17)	9.96	(2.77, 2.91)	8.28
(2.17, 2.19)	9.72	(2.91, 3)	8.52

is $T = 7.80$. Tables 2, 3 and 4 shows the optimal maintenance decisions for the different values of δ, α and β, respectively.

The values of the PEVPI's along with the uncertainty intervals are given in Table 5. From these results, it can be concluded that η and δ (the shape and scale parameters of the lifetime distribution) are the most important factors. Note that knowing η prior to the decision shows the most substantial differentiation between optimal strategies. Thus, this parameter is 'important' in the sense that reducing uncertainty about its value is likely to result in a different optimal strategy. This conclusion is further supported in Fig. 3 which summaries the sensitivity analysis of the cost rate

Table 5 The PEVPI's estimated based on the fitted GP process for the parameters of the RV model for the TBM policy

θ_i	$PEVPI_i$	$C.I$
η	0.109	(0.102, 0.116)
δ	0.0268	(0.0199, 0.0338)
α	0.0075	(0.0235, 0.0213)
β	0.009	(0.00216, 0.0159)

Fig. 2 The variance contribution of each input parameter to the cost rate function of the time-based maintenance policy at $T = 8.5$ for the RV deterioration model

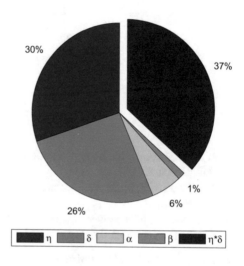

function with respect to the changes of the model input parameters. In this Fig. 2, the variance contribution of each parameter to the total variance of the cost rate at $T = 8.5$ is shown. The variance contribution of η and δ is almost 30% and 26% based on only 60 data-points at T = 8.5, respectively, while α and β covers only 6% and 1% of total variances, respectively. In addition to the individual impacts of η and δ, their interaction which covers 37% of total variance is considered as an important factor influencing the maintenance cost at the chosen optimal replacement time. The analysis also exposes the behaviour of the expected cost at a specific time for different values of the parameters. Figure 3 illustrates how expected cost $E_{\theta|\theta_i}\left[-\mathscr{C}_R(t;\theta)\right]$ when T = 8.5 changes with different values of the parameters (i.e. $(\eta, \delta, \alpha, \beta)$), along a 95% uncertainty bound (the thickness of the band).

4.2 The Condition-Based Maintenance Policy

Under the RV deterioration model, the following CBM policy is considered, based on the periodic inspection of a component at a fixed time interval t_I with cost, C_I (see also [22])

Fig. 3 Expected utilities
and 95% uncertainty bounds
for $T = 8.5$ when the
parameters are completely
known before the
maintenance decision

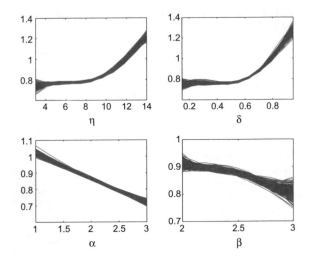

1. If $X(t_I) \leq \upsilon\rho$, no maintenance action should be taken until the next inspection with the cost, C_I.
2. If $\upsilon\rho < X(t_I) \leq \rho$, the maintenance action will be taken at the cost C_P to renew the system.
3. If the system fails between inspections ($X(t_I) > \rho$), a corrective maintenance will restore the system at the cost of C_F

where $0 < \upsilon < 1$ is called PM ratio, and $\upsilon\rho$ is the threshold for the PM which is a fraction of the failure threshold.

According to renewal theory [2, 23], the mean cost rate for the CBM policy under the RV deterioration model is given by

$$\mathscr{K}_R(t_I, \upsilon; \theta) = \frac{E[\mathscr{C}_U(t_I, \upsilon; \theta)]}{E[\mathscr{L}_D(t_I, \upsilon; \theta)] + \tau_r}, \tag{25}$$

where

$$E[\mathscr{C}_U(t_I, \upsilon; \theta)] = (C_I + C_P)Pr(X(t_I) \leq \rho) + C_F Pr(X(t_I) > \rho)$$
$$= (C_I + C_P - C_F)\mathscr{G}(\rho/t_I; \eta, \delta) + C_F,$$

$$E[\mathscr{L}_D(t_I, \upsilon; \theta)] = \int_0^{t_I} Pr(X(t) < \rho)dt + \int_{t_I}^{\infty} Pr(X(t) < \upsilon\rho)dt$$
$$= \int_0^{t_I} \mathscr{G}(\rho/t; \eta, \delta)dt + \int_{t_I}^{\infty} \mathscr{G}(\upsilon\rho/t; \eta, \delta)dt,$$

τ_r is the expected duration of the maintenance action as given in (23), and $\theta = (\eta, \delta, \alpha, \beta)$.

In [22, 23], it was discussed that the optimal inspection time (t_I) is unique and will lie in an interval with the details given in these works (e.g. $t_I \in [3, 16]$ in the following numerical example). The optimal inspection interval and PM ratio, (t_I, v) apparently depends on the parameter values, θ. Thus, the robust optimal inspection interval can be derived through performing a proper decision-based sensitivity analysis described above.

The PM ratio, v is considered as an extra parameter and included into the uncertain parameters, that is, $\phi = (\eta, \delta, \alpha, \beta, v)$. The joint prior distribution given in (24) will be revised as follows:

$$\pi(\phi) = \pi_1(\eta)\pi_2(\delta)\pi_3(\alpha)\pi_4(\beta)\pi_5(v), \tag{26}$$

where

$$\eta \sim U(3, 14), \quad \delta \sim U(0.15, 0.95), \quad \alpha \sim U(1, 3), \quad \beta \sim U(2, 3), \quad v \sim U(0, 1)$$

To compute the PEVPI's associated with these input parameters, the TI emulator is then fitted to the mean cost rate, $\mathcal{K}_R(t_I, \phi)$ based on 2160 training data points. These data points consists of 80 design points generated from the joint prior distribution of ϕ (using the Latin hypercube design) and then evaluating the mean cost rate at each of these design points for 27 selected values for t_I, that is, $t_I = 3, 3.5, \ldots, 16$.

The maximum benefit of using the information available in $\pi(\phi)$ is $U_0 = 0.642$ which attains at the optimal inspection time interval $T = 6.62$. Tables 8, 9, 10 and 11 in Appendix A show the optimal inspection interval decisions when the values of γ, ξ, α and β are learned prior to making any decision about the inspection time. Table 6 illustrates the optimal t_I when the values of PM ration v is learned prior to the decision-making. For instance, when the PM ratio lies in $(0.625, 0.925)$, then the mean cost rate is minimum at $t_I = 5.1$, but when v takes values closer to its median, that is, $v \in (0.475, 0.525)$, then $t_I = 5.65$.

The values of the PEVPI's along with their 95% confidence bounds over the considered interval for optimal inspection time (i.e. $I_2 = [3, 16]$) are presented in Table 7. The parameters with the highest PEVPI's are δ and η (the scale and shape parameters

Table 6 Optimal decisions when PM ratio parameter, v is known prior the maintenance decision

Range	T	Range	T
(0, 0.025)	8.95	(0.375, 0.475)	6.20
(0.025, 0.075)	8.40	(0.475, 0.525)	5.65
(0.075, 0.175)	7.85	(0.525, 0.625)	4
(0.175, 0.275)	7.30	(0.625, 0.925)	5.10
(0.275, 0.375)	6.75	(0.925, 1)	4

Table 7 The estimated PEVPI's based on the fitted GP emulator for the parameters of the RV model for the CBM policy

θ_i	$PEVPI_i$	$C.I$
η	0.0117	(0.005, 0.018)
δ	0.0174	(0.01, 0.0243)
α	0.008255	(0.0014, 0.015)
β	0.0085	(0.0017, 0.015)
υ	0.01184	(0.004, 0.02)

Fig. 4 The variance contribution of each input parameters to the cost rate function of the CBM policy at $t_I = 6.6$ for the RV deterioration model

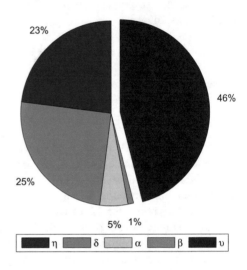

of the RV deterioration model) and the PM ratio parameter, υ. Learning about these parameters prior to making when to inspect the component of interest, will allow the decision-maker to implement a strategy that will result in the higher cost saving. That also means these parameters are essential in the sense that, reducing uncertainty about their values, is likely to change the optimal inspection interval.

Figure 4 also illustrates the sensitivity analysis of $\mathscr{K}_R(t_I, \phi)$ with respect to the changes of ϕ at $t_I = 6.6$. Based on the variance contribution fractions, it can be similarly concluded that υ, δ and η (23%) are the most influencing factors affecting the mean cost rate at $t_I = 6.6$.

5 Conclusions and Discussion

Strategic planning for information systems implementation requires a thorough assessment of the to-be system's lifecycle, resilience and overall sustainability. This is of no exception for cloud services and big data analytic systems. The overall aim is to reduce the energy consumption of the system, enhance system's environmental friendliness, align the system with social aspects of sustainability, reduce the cost, prolong system's endurance or a combination of all of the above. There are numerous facets to information systems strategic engineering. Hosseinian-Far and Chang [13] assessed the information system's sustainability, e.g. Strategic Information Systems (SIS)'s endurance by quantifying seven factors, two of which were contextualised for the specific SIS's under study. However, the research presented in this paper, surpasses that by focusing on the preventive maintenance strategies by which the system's lifecycle can be prolonged. This is in line with the economy domain of the three pillar generic sustainability model i.e. People, Planet and Profit [12], as it entails less costs as opposed to post issue maintenance strategies, i.e. Corrective Maintenance (CM) strategies.

In this chapter, we have demonstrated how the life-cycle management of an asset (system) under continuous deterioration can be efficiently and effectively improved by studying the sensitivity of optimised PM decisions with respect to changes in the model parameters. The novelty of this research is the development of a computationally efficient decision-theoretic sensitivity analysis for a PM optimisation problem for infrastructure assets subject to measurable deterioration using the PEVPI concept. This approach would enable the decision-maker to select a PM strategy amongst an infinite set of decisions in terms of expected costs and in terms of accumulated information of the model parameters and aspects of the system, including deterioration process and maintenance models. We use an RV deterioration model to present the uncertainty associated with asset deterioration across both age-based and condition-based PM policies. The computation of PEVPI's is very demanding, and conventional numerical integration or Monte Carlo simulation techniques would not be as helpful. To overcome this computational challenge, we introduced a new approach which approximates the PEVPI using GP emulators; a computationally efficient method for highly complex models which require fewer model runs than other approaches (including MCMC based methods). The method is illustrated on worked numerical examples and discussed in the context of analytical efficiency and, importantly, organisational learning.

The EVPI-based sensitivity analysis presented here can be used for other maintenance optimisation problems including problems with imperfect maintenance [24], or delay-time maintenance [4]. In this case, it is considered as one of the more effective preventive maintenance policies for optimising inspection planning. An efficient condition-based maintenance strategy which allows us to prevent system/component failure by detecting the defects via an optimised inspection might be identified using the sensitivity analysis proposed in this paper to determine a robust optimal solution for delay-time maintenance problems and the expected related costs, when the cost function parameters are either unknown or partially known.

Appendix A: Optimal CBM Maintenance Decisions' Tables

See Tables 8, 9, 10 and 11.

Table 8 Optimal decisions when parameter η is known prior the maintenance decision

Range	t_I	Range	t_I
(3, 3.335)	10.97	(7.585, 7.95)	9.13
(3.335, 3.7)	12.07	(7.95, 8.315)	8.77
(3.70, 4.385)	12.80	(8.315, 8.585)	8.4
(4.385, 5.015)	12.43	(8.5855, 8.95)	8.03
(5.015, 5.385)	11.70	(8.95, 9.415)	7.67
(5.385, 5.75)	11.3	(9.415, 9.785)	7.3
(5.75, 6.115)	10.97	(9.785, 10.15)	6.93
(6.115, 6.485)	10.6	(10.15, 10.885)	6.57
(6.485, 6.85)	12.23	(10.885, 11.615)	6.20
(6.485, 7.215)	9.9	(11.615, 14)	5.83
(7.215, 7.585)	9.5		

Table 9 Optimal decisions when parameter δ is known prior the maintenance decision

Range	t_I	Range	t_I
(0.15, 0.1758)	11.33	(0.457, 0.483)	9.87
(0.1758, 0.19)	15	(0.483, 0.51)	9.5
(0.19, 0.217)	14.63	(0.51, 0.537)	8.77
(0.217, 0.243)	13.17	(0.537, 0.563)	8.40
(0.243, 0.297)	12.80	(0.563, 0.59)	8.03
(0.297, 0.323)	12.43	(0.59, 0.617)	7.77
(0.323, 0.35)	12.07	(0.617, 0.643)	7.30
(0.35, 0.377)	11.70	(0.643, 0.67)	6.93
(0.377, 0.403)	11.33	(0.67, 0.723)	6.57
(0.403, 0.43)	10.97	(0.723, 0.803)	6.20
(0.43, 0.457)	10.23	(0.803, 0.937)	5.83
		(0.937, 0.95)	5.47

Table 10 Optimal decisions when parameter α is known prior the maintenance decision

Range	t_I
(1, 3)	6.64

Table 11 Optimal decisions when parameter β is known prior the maintenance decision

Range	t_I
$(2, 2.15) \cup (2.83, 2.95)$	6.42
$(2.15, 2.41) \cup (2.59, 2.83)$	6.64
$(2.41, 2.59)$	6.86
$(2.95, 3)$	6.20

References

1. Ahmad, R., & Kamaruddin, S. (2012). An overview of time-based and condition-based maintenance in industrial application. *Computers & Industrial Engineering, 63*, 135149.
2. Aven, T., & Dekker, R. (1997). A useful framework for optimal replacement models. *Reliability Engineering & System Safety, 58*, 61–67.
3. Barlow, R. E., Lai, C. D., & Xie, M. (2006). *Stochastic ageing and dependence for reliability.* New York: Springer Science & Business.
4. Christer, A. H., & Wang, W. (1995). A delay-time-based maintenance model of a multi-component system. *IMA Journal of Mathematics Applied in Business & Industry, 6*, 205–222.
5. Conti, S., & O'Hagan, A. (2010). Bayesian emulation of complex multi-output and dynamic computer models. *Journal of Statistical Planning and Inference, 140*, 640–651.
6. Daneshkhah, A., & Bedford, T. (2013). Probabilistic sensitivity analysis of system availability using Gaussian processes. *Reliability Engineering and System Safety, 112*, 8293.
7. De Jonge, B., Dijkstra, A. S., & Romeijnders, W. (2015). Cost benefits of postponing time-based maintenance under lifetime distribution uncertainty. *Reliability Engineering & System Safety, 140*, 1521.
8. Dekker, R., & Plasmeijer, R. P. (2001). Multi-oparameter maintenance optimisation via the marginal cost approach. *Journal of the Operational Research Society, 52*, 188–197.
9. Dekker, R. (1996). Applications of maintenance optimization models: a review and analysis. *Reliability Engineering & System Safety, 51*, 229–240.
10. Dobbs, R., Pohl, H., Lin, D., Mischke, J., Garemo, N., Hexter, J., et al. (2013). *Infrastructure productivity: How to save $1 trillion a year.* McKinsey Global Institute.
11. Ellis, H., Jiang, M., & Corotis, R. (1995). Inspection, maintenance and repair with partial observability. *Journal of Infrastructure Systems, 1*(2), 9299.
12. Elkington, J. (1994). Towards the sustainable corporation: Win-Win-Win business strategies for sustainable development. *California Management Review, 36*(2), 90–100.
13. Hosseinian-Far, A., & Chang, V. (2015). Sustainability of strategic information systems in emergent vs. prescriptive strategic. *Management, International Journal of Organizational and Collective Intelligence, 5*(4), 1–7.
14. Kennedy, M. C., & O'Hagan, A. (2001). Bayesian calibration of Computer models (with discussion). *Journal of the Royal Statistical Society: Series B, 63*, 425–464.
15. Kleijnen, J. P. C. (2009). Kriging metamodeling in simulation: A review. *European Journal of Operational Research, 192*, 707–716.
16. McKay, M. D., Beckman, R. J., & Conover, W. J. (1979). A comparison of three methods for selecting values of input variables in the analysis of output from a computer code. *Technometrics, 21*(2), 239–245.
17. Nicolai, R. P., & Dekker, R. (2007). A review of multi-component maintenance models. In T. Aven & J. E. Vinnem (Eds.), *Risk, Reliability and Societal Policy, VOLS 1–3—VOL 1: Specialisation topics; VOL 2: Thematic topics; VOL 3: Application topics* (Vol. 1, pp. 289–296). London, England: Taylor & Francis LTD.
18. O'Hagan, A. (2006). Bayesian analysis of computer code outputs: A tutorial. *Reliability Engineering & System Safety, 91*, 1290–1300.

19. O'Hagan, A. (1991). Bayes-Hermit quadrature. *Journal of Statistical Planning & Inference*, *91*, 245–260.
20. Oakley, J., & O'Hagan, A. (2004). Probabilistic sensitivity analysis of complex models: A bayesian approach. *Journal of the Royal Statistical Society Series B*, *66*, 751–769.
21. Oakley, J. E. (2009). Decision-theoretic sensitivity analysis for complex computer models. *Technometrics*, *51*(2), 121–129.
22. Pandey, M. D., Yuan, X. X., & Van Noortwijk, J. M. (2009). The influence of temporal uncertainty of deterioration on life-cycle management of structures. *Structure and Infrastructure Engineering*, *5*(2), 145156.
23. Park, K. S. (1988). Optimal continuous-wear limit replacement under periodic inspections. *IEEE Transactions on Reliability*, *37*(1), 97102.
24. Pham, H., & Wang, H. (1996). Imperfect maintenance. *European Journal of Operational Research*, *94*(3), 425–438.
25. Rougier, J. (2008). Efficient emulators for multivariate deterministic functions. *Journal of Computational and Graphical Statistics*, *14*(4), 827–843.
26. Sacks, J., Welch, W. J., Mitchell, T. J., & Wynn, H. P. (1989). Design and analysis of computer experiments. *Statistical Science*, *4*(4), 40935.
27. Saltelli, A. (2002). Sensitivity analysis for importance assessment. *Risk analysis*, *22*(3), 579–590.
28. Wang, H. Z. (2002). A survey of maintenance policies of deteriorating systems. *European Journal of Operational Research*, *139*(3), 469–489.
29. Zitrou, A., Bedford, T., & Daneshkhah, A. (2013). Robustness of maintenance decisions: Uncertainty modelling and value of information. *Reliability Engineering & System Safety*, *120*, 60–71.

A Novel Safety Metric SM_{EP} for Performance Distribution Analysis in Software System

R. Selvarani and R. Bharathi

Abstract Focusing on safety attributes becomes an essential practice towards the safety critical software system (SCSS) development. The system should be error free for a perfect decision-making and subsequent operations. This paper presents an analysis on error propagation in the modules through a novel safety metric known as SM_{EP}, which can be characterized depending on the performance rate of the working module. We propose a framework for the analysis of occurrence of error in various modules and the intensity of it is quantified through probabilistic model and universal generating function technique.

Keywords Safety critical · Error propagation · UGF · Performance · Modules

1 Introduction

SCSS are pervasive in the medical field and it has to be designed with maximum care. The dependability requirement is an important criterion in these systems. To reduce the probability of losses, appropriate failure analysis practices have to be used to validate the safety of the critical system [1].

To achieve error free scenario of SCSS is very difficult, although the system has been well tested, used, and documented. If one part of a system fails, this can affect other parts and in worst case results in partial or even total system failure. To avoid such incidents, research on failure analysis is of high importance. Failure analysis is the proven approach for mitigating the system hazards and failure modes and consequently determines which of those are influenced by or affected by software or

R. Selvarani
Computer Science and Engineering, Alliance University, Banglore, India
e-mail: selvarani.r@alliance.edu.in

R. Bharathi (✉)
Information Science and Engineering, PESIT-BSC, Banglore, India
e-mail: rbharathi@pes.edu

© Springer International Publishing AG 2017
A. Hosseinian-Far et al. (eds.), *Strategic Engineering for Cloud Computing and Big Data Analytics*, DOI 10.1007/978-3-319-52491-7_3

lack of software [1]. A failure of a safety critical system can be defined as "the non performance or incapability of the system or a component of the system to perform or meet the expectation for a specific time under stated environmental conditions."

The error propagation probability is a condition that once an error occurs in a system module, it might propagate to other modules and thereby cascades the error to the system output [2]. The error propagation analysis is a vital activity for the efficient and robust designing of safety critical software system.

Error propagation between software modules is a quantitative factor that reflects on the reliability of a safety critical software product. In general, the SCSS is considered between two major states, perfect functioning and failure state. Here we are considering several intermittent states between the two major states for the failure analysis. Hence these systems can be termed as *Multistate Systems* (MS) in our research. The reliability of a MS can be defined as a measure of the capability of the system to execute required performance level [3].

The presence of an error [4] in a software module might trigger an error in other modules of the system that are interconnected. Identifying error propagation in a software system is an important task during the development activity. Propagation analysis may be used to identify the critical modules in a system, and to determine how other modules are affected in the presence of errors. This concept will aid in system testing and debugging through generating required test cases that will stimulate fault activation in the identified critical modules and facilitate error detection [5].

The errors under consideration might be due to faulty design, which could result in errors and data errors due to wrong data, late data, or early data. The impact of error propagation across modules can be assessed by analyzing the error propagation process and arrive at a general expression to estimate the performance distribution of each module using computational intelligence because of its complexity and randomness [6]. As per IEC 61508, it is necessary to see that the design and performance of critical systems is safety enough to meet tolerable risk targets, taking into account of all failure sources including systematic hardware and software faults and random faults [7].

The reliability and performance of a multistate safety critical system can be computed by using Universal Generating Function (UGF) technique [8]. The UGF technique is based on probability theory to assess and express models through polynomial functions. The UGF technique applied for failure analysis in safety critical systems in this paper is adapted by following the procedure given by Levitin et al. [8, 9].

Hence the error propagation analysis provides the base for the reliability evaluation, since the occurrence of error propagation across the modules has a significant effect on the system behavior during critical states.

The paper is structured as follows: Sect. 2 describes background and Sect. 3 describes the proposed approach through a framework. The analysis of error propagation and failure of a SCSS is depicted in Sect. 4. Conclusion and looking beyond the area of this research are discussed in Sect. 5.

2 Background

According to Avizienis et al. [4], a *failure* is an event that occurs when the delivered service no longer complies with the expected service of the system. An *error* is an incorrect internal state that is liable to the occurrence of a failure or another error. However all errors may not reach the system's output to cause a failure. A *fault* is active when it results in an error otherwise it is said to be inactive. Nevertheless, not all faults lead to an error, and not all errors lead to a failure.

2.1 Error Propagation

Error propagation (EP) can be defined as a condition where a failure of a component may potentially impact other system components or the environment through interactions, such as communication of bad data, no data, and early/late data [10]. Such failures represent hazards that if not handled properly can result in potential damage. The interaction topology and hierarchical structure of the system architecture model provide the information on how the erroneous behavior of each system component interacts through error propagation.

Morozov et al. [5] have used probabilistic error propagation techniques for diagnosing the system. Henceforth it aids in tracing back the path of error propagation path to the error-origin. Moreover this diagnosis helps in error localization procedure, testing, and debugging [5].

The influence of error propagation in the overall system reliability has been demonstrated in [11]. With the help of UML artifacts, the system architectural information is used to find the probability of error propagation across system components [11]. Since they have used UML artifacts, their model can be used to predict reliability in the early phases of system development.

Hiller et al. [12] have initiated a new concept called "Error Permeability" through software modules, as well as a set of associated measures namely error exposure, relative permeability, and non-weighted relative permeability. They found these measures are helpful in assessing the vulnerable modules, which are exposed to error propagation.

A bottom-up approach is considered to estimate the reliability for component-based system in [2]. Foremost, the reliability of system component was assessed. Based on the architectural information, the system reliability was estimated taking into the account of error propagation probability. The system analysis was carried out through the failure model by considering only data errors across components. Authors in [2] have concluded that error propagation is a significant characteristic of each system component and defined as the probability that a component propagates the erroneous inputs to the generated output. Their approach can be used in the early prediction of system reliability.

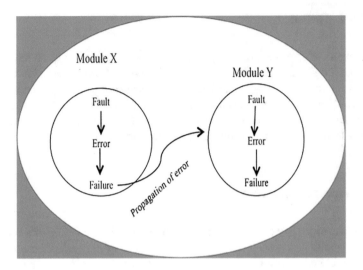

Fig. 1 Inter-modular error propagation

An error happens when there is an activation of a fault [4]. An error occurs in a module when there is a fault in the module and henceforth it cannot directly cause an error in other modules. Relatedly an error in a module can lead to its failure only within that module. The reason for the module error is either due to the activation of fault in the same module or due to deviated input service from other modules. A module failure is defined as the deviation of the performance from its accepted output behavior. If the failed module is the output interface module of the system then its failure is considered as a system failure [13]. System failures are defined based on its boundary. A system failure is said to occur when error propagates outside the system. Figure 1 depicts intermodular error propagation. Module X influences module Y.

In safety critical system, certain factors are considered crucial which signifies the safety of a system and such critical attributes should be consistently monitored throughout the lifecycle of the system [14]. This work focuses on analyzing error propagation in safety critical software systems. In this approach, we use a methodology called universal generating function (UGF) to quantify the performance distribution of a multistate safety critical software system [3] and subsequently introduce a new metric called Safety Metric (SM_{EP}).

2.2   MSS and Universal Generating Function

The UGF technique also called as u function is a mathematical tool introduced by Ushakov [15] and Levitin [8] expanded and proved that UGF is an effective technique for assessing the performance of real-world systems, in specific

Multistate Systems. In general all traditional reliability models perceived system as binary state systems, states being a perfect functionality and a complete failure. In reality, each system has different performance levels and various failure modes affecting the system performance [3]. Such systems are termed as Multistate Systems (MS).

Let us assume a MS composed of n modules. In order to assess the reliability of a MS, it is necessary to analyze the characteristic of each module present in the system. A system module 'm' can have different performance rates and represented by a finite set q_m, such that $q_m = \{q_{m1}, q_{m2}, \ldots q_{mi} \ldots q_{mk_m}\}$ [16], where q_{mi} is the performance rate of module m in the ith state and $q_i = \{1, 2, \ldots k_m\}$. The performance rate $Q_m(t)$ of module 'm', at time $t \geq 0$ is a random variable that takes its value from $q_m: Q_m(t) \in q_m$.

Let the ordered set $p_m = \{p_{m1}, p_{m2}, \ldots p_{mi}, \ldots p_{mj_m}\}$ associate the probability of each state with performance rate of the system module m, where $p_{mi} = Pr\{Q_m = q_{mi}\}$.

The mapping $q_{mi} \rightarrow p_{mi}$ is called the probability mass function (pmf) [17].

The random performance [18] of each module m defined as polynomials can be termed as module's UGF $(u_m(z))$

$$u_m(z) = \sum_{i=0}^{k} P_{mi} z^{q_{mi}}, \text{ where } m = 1, 2 \ldots n. \tag{1}$$

Similarly the performance rates of all 'm' system modules have to be determined. At each instant $t \geq 0$, all the system modules have their performance rates corresponding to their states. The UGF for the MS denoted as "$(U_S(Z))$" can be arrived, by determining the modules interconnection through system architecture. The random performance of the system as a whole at an instant $t \geq 0$ is dependent on the performance state of its modules. The UGF technique specifies an algebraic procedure to calculate the performance distribution of the entire MS, denoted as $U_S(z)$,

$$U_s(z) = f\{u_{m1}(z), u_{m2}(z), \ldots, u_{mn}(z)\}, \tag{2}$$

$$U_s(z) = \nabla_\phi \{u_{m1}(z), u_{m2}(z), \ldots, u_{mn}(z)\} \tag{3}$$

where ∇ is the composition operator and ϕ is the system structure function. In order to assess the performance distribution of the complete system with the arbitrary structure function ϕ, a composition operator ∇ is used across individual u *function* of m system modules [17].

$U_S(z)$ is a U function representation of performance distribution of the whole MS software system. The composition operator ∇ determines the U function of the whole system by exercising numerical operations on the individual u functions of the system modules. The structure function $\phi(\cdot)$ in composition operator ∇

expresses the complete performance rate of the system consisting of different modules in terms of individual performance rates of modules. The structure function $\phi(\cdot)$ depends upon the system architecture and nature of interaction among system modules.

Reliability is nothing but continuity of expected service [4] and it is well known that, it can be quantitatively measured as failures over time. The UGF technique can be used for estimating software reliability of the system as a whole consisting of n modules. Each of the modules performs a sub-function and the combined execution of all modules performs a major function [17].

An assumption while using the UGF technique is that the system modules are mutually independent of their performance.

3   Proposed Approach

Error Propagation (EP) is defined as the condition where an error (or failure) propagates across system modules [19]. Our approach focuses on quantifying the propagation of error between modules in safety critical software system.

The analysis proposed in this research contains four different stages and explained through a framework. The framework, as shown in Fig. 2, is based on bottom-up approach in assessing the performance distribution of SCSS To start with, we have arrived at the performance distribution of each system module (PD_{MOD}) using U *function*.

The probability of error propagation in a module ($PD_{MOD} + SM_{EP}$) is quantified in the second step. As third step, the performance distribution of subsystems ($PD_{SS} + SM_{EP}$) is arrived through composition operator. As the final step the failure prediction is achieved through recursive operations for quantifying the error propagation throughout the system ($PD_{SYS} + SM_{EP}$).

During software development, this framework would be helpful to demonstrate the probability of error propagation to identify the error prone areas.

4   Error Propagation and Failure Analysis

The error propagation and failure analysis model is a conceptual framework for analyzing the occurrence of error propagation in SCSS. The system considered is broken down into subsystem, and each subsystem in turn is subdivided into modules called elements.

A module is an atomic structure, which performs definite function(s) of a complex system.

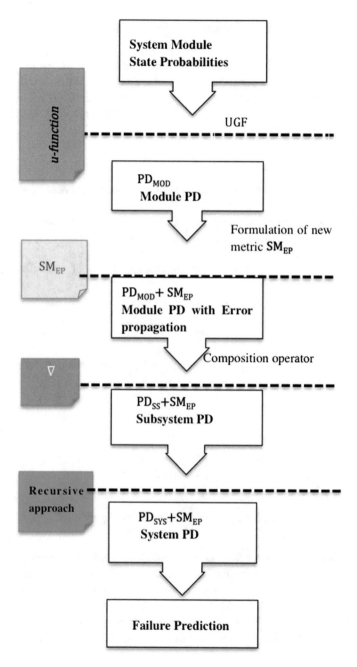

Fig. 2 Error propagation and failure analysis

4.1 Performance Distribution of System Module

The performance rate of a module can be measured in terms of levels of failure.

Let us assume that the performance rate of a module m with 0% failure is q_{m1}, 10% failure is q_{m2}, 30% failure is q_{m3}, 50% failure is q_{m4}, and 100% failure is q_{m5}.

The state of each module m can be represented by a discrete random variable Q_m that takes value from the set,

$$Q_m = \{q_{m1}, q_{m2}, q_{m3}, q_{m4}, q_{m5}\}$$

The random performance of a module varies from perfect functioning state to complete failure state.

The probabilities associated with different states (performance rates) of a module m at time t can be represented by the set, $P_m = \{p_{m1}, p_{m2}, p_{m3}, p_{m4}, p_{m5}\}$, where $P_{mh} = Pr\{Q_m = q_{mh}\}$.

The module's states is the composition of the group of mutually exclusive events,

$$\sum_{h=1}^{5} P_{mh} = 1 \tag{5}$$

The performance distribution of a module m (pmf of discrete random variable G) can be defined as

$$u_m(z) = \sum_{h=1}^{5} P_{mh} z^{q_{mh}} \tag{6}$$

The performance distribution of any pair of system modules l and m, connected in series or parallel [18], can be determined by,

$$u_l(z) \nabla u_m(z) = \sum_{h=1}^{5} P_{lh} z^{q_{ih}} \nabla \sum_{h=1}^{5} P_{mh} z^{q_{mh}} \tag{7}$$

The composition operator ∇ determines the *u function* for two modules based on whether they are connected in parallel or series using the structure function ø. The equation arrived in Eq. (7) quantifies the performance distribution of combination of modules. Levitin et al. in [9] have demonstrated the determination of performance distribution when the modules are connected in series or parallel. A failure in a module may potentially impact other modules through error propagation.

4.2 Formulation of Safety Metric SM$_{EP}$

The probability of occurrence of EP in a module can be defined by introducing a new state in that module [9]. Assuming that the state 0 of each module corresponds to the EP originated from this module [8]. The Eq. (6) can be rewritten as,

$$u_m(z)_{ep} = P_{m0}z^{q_{m0}} + \sum_{h=1}^{5} P_{mh}z^{q_{mh}} \tag{8}$$

$$u_m(z)_{ep} = P_{m0}z^{q_{m0}} + u_m(z) \tag{9}$$

where p_{m0} is the probability state for error propagation and q_{m0} is the performance of the module at state 0.

$u_m(z)$ represents all states except the state of error propagation as represented in Eq. (6).

The performance distribution of a module m at state 0 is the state of error propagation which is given by $P_{m0}z^{q_{m0}}$ and is termed as safety metric SM$_{EP}$. This metric depends upon the probability of the module performance with respect to EP, whether a failed module can cause EP or not and whether a module can be infuenced by EP or not. The safety metric SM$_{EP}$ of each module will carry a weightage based on the probability of propagating error. If the module does not propagate any error, the corresponding state probability should be equated to zero [9].

$$p_{m0} = 0 \tag{10}$$

By substituting Eq. (10) in Eq. (8), the SM$_{EP}$ is quantified as zero. Therefore Eq. (8) becomes,

$$u_m(z)_{ep} = \sum_{h=1}^{5} P_{mh}z^{q_{mh}} \tag{11}$$

The module that does not have error propagation property or state is given by $u_m(z)_{ep} = u_m(z)$, the u function in Eq. (11) is reduced to Eq. (6).

If a failed module causes error propagation, then the performance of the module in that state of error propagation is

$$q_{m0} = \alpha \tag{12}$$

where the value of α can be of any random performance q_{m1} or q_{m2} or q_{m3} or q_{m4} or q_{m5}. When any operational module m that will not fail due to error propagation can be represented by conditional pmf [9],

$$u_m(z)_{ep} = \sum_{h=1}^{5} \frac{p_{mh}}{1 - p_{m0}} z^{q_{mh}} \tag{13}$$

Because the module can be in any one of the five states as defined in Eq. (6). The safety metric SM_{EP} depends on the performance of each module in the multistate system. This metric helps to estimate the probability of EP to hazardous modules of the system and identify modules that should be protected with fault detection and fault recovery mechanisms.

Based on the above considerations, the conditional u functions of each system module have to be estimated. Depending upon the subsystem architecture, the u function of each subsystem can be quantified by applying the composition operator ∇_ϕ. Then the recursive approach is used to obtain the entire *u function* of safety critical software system, which will be elaborated in the subsequent work.

5 Conclusions

This approach proposes a new framework to analyze the failure of multistate safety critical software with respect to error propagation and arrive at a new metric called safety metric SM_{EP}. This proposed new metric will be the key finding for the failure analysis of real-time safety critical system. This metric has the application in finding the failure probability of each module, the migration of error propagation from modular level to subsystem and then to system level and the process of identifying the most critical module in the whole safety critical software system and the impact of error propagation in the performance of SCSS. Our future work will continue by applying the safety metric SM_{EP} in relevant real-time SCSS for its failure analysis.

References

1. Sundararajan, A., & Selvarani, R. (2012). Case study of failure analysis techniques for safety critical systems. In *Advances in Computer Science, Engineering & Applications* (pp. 367–377). Heidelberg: Springer.
2. Cortellessa, V., & Grassi V. (2007). A modeling approach to analyze the impact of error propagation on reliability of component-based systems. In *International Symposium on Component-Based Software Engineering*. Heidelberg: Springer.
3. Levitin, G. (2008). A universal generating function in the analysis of multi-state systems. In *Handbook of performability engineering* (pp. 447–464). London: Springer.
4. Avizienis, A., et al. (2004). Basic concepts and taxonomy of dependable and secure computing. *IEEE Transactions on Dependable and Secure Computing 1.1*, 11–33.
5. Morozov, A., & Janschek, K. (2014). Probabilistic error propagation model for mechatronic systems. *Mechatronics, 24*(8), 1189–1202.
6. Eberhart, R. C., & Shi, Y. (2007). Chapter nine—Computational intelligence implementations. In *Computational intelligence* (pp. 373–388). Burlington: Morgan Kaufmann. ISBN 9781558607590.
7. IEC 61508. (2005). Functional Safety of Electrical/Electronic/Programmable Electronic Safety-related Systems. International Electro technical Commission (IEC 61508).
8. Levitin, G. (2007). Block diagram method for analyzing multi-state systems with uncovered failures. *Reliability Engineering & System Safety, 92*(6), 727–734.

9. Levitin, G., & Xing, L. (2010). Reliability and performance of multi-state systems with propagated failures having selective effect. *Reliability Engineering & System Safety, 95*(6), 655–661.
10. Feiler, P. H., Goodenough, J. B., Gurfinkel, A., Weinstock, C. B., Wrage, L. (2012). Reliability validation and improvement framework. Carnegie-Mellon Univ Pittsburgh Pa Software Engineering Inst.
11. Popic, P., et al. (2005). Error propagation in the reliability analysis of component based systems. In *16th IEEE International Symposium on Software Reliability Engineering (ISSRE'05)*. IEEE.
12. Hiller, M., Jhumka, A., & Suri, N. (2001). An approach for analysing the propagation of data errors in software. In *International Conference on Dependable Systems and Networks, 2001, DSN 2001*. IEEE.
13. Mohamed, A., & Zulkernine M. (2010). Failure type-aware reliability assessment with component failure dependency. In *2010 Fourth International Conference on Secure Software Integration and Reliability Improvement (SSIRI)*. IEEE.
14. Vinisha, F. A., & Selvarani, R. (2012). Study of architectural design patterns in concurrence with analysis of design pattern in safety critical systems. In *Advances in Computer Science, Engineering & Applications* (pp. 393–402). Heidelberg: Springer.
15. Ushakov, I. A. (1986). A universal generating function. *Soviet Journal of Computer and Systems Sciences, 24*(5), 118–129.
16. Levitin, G., & Lisnianski, A. (2003). Multi-state system reliability: Assessment, optimization and applications.
17. Levitin, G. (2005). *The universal generating function in reliability analysis and optimization.* London: Springer.
18. Levitin, G., Zhang, T., & Xie, M. (2006). State probability of a series-parallel repairable system with two-types of failure states. *International Journal of Systems Science, 37*(14), 1011–1020.
19. Sarshar, S. (2011). Analysis of Error Propagation Between Software Processes. *Nuclear Power-System Simulations and Operation*, 69.

Prior Elicitation and Evaluation of Imprecise Judgements for Bayesian Analysis of System Reliability

Alireza Daneshkhah, Amin Hosseinian-Far, Tabassom Sedighi and Maryam Farsi

Abstract System reliability assessment is a critical task for design engineers. Identifying the least reliable components within a to-be system would immensely assist the engineers to improve designs. This represents a pertinent example of data-informed decision-making (DIDM). In this chapter, we have looked into the theoretical frameworks and the underlying structure of system reliability assessment using prior elicitation and analysis of imprecise judgements. We consider the issue of imprecision in the expert's probability assessments. We particularly examine how imprecise assessments would lead to uncertainty. It is crucial to investigate and assess this uncertainty. Such an assessment would lead to a more realistic representation of the expert's beliefs, and would avoid artificially precise inferences. In other words, in many of the existing elicitation methods, it cannot be claimed that the resulting distribution perfectly describes the expert's beliefs. In this paper, we examine suitable ways of modelling the imprecision in the expert's probability assessments. We would also discuss the level of uncertainty that we might have about an expert's density function following an elicitation. Our method to elicit an expert's density function is nonparametric (using Gaussian Process emulators), as introduced by Oakley and O'Hagan [1]. We will modify this method by including the imprecision in any elicited probability judgement. It should be noticed that modelling imprecision does not have any impact on the expert's true density function, and it only affects the analyst's uncertainty about the unknown quantity of interest. We will compare our

A. Daneshkhah (✉)
School of Engineering, The Warwick Centre for Predictive Modelling,
The University of Warwick, Coventry CV4 7AL, UK
e-mail: ardaneshkhah@gmail.com; a.daneshkhah@warwick.ac.uk

A. Hosseinian-Far
School of Computing, Creative Technologies & Engineering, Leeds Beckett University,
Leeds LS6 3QR, UK
e-mail: A.Hosseinian-Far@leedsbeckett.ac.uk

T. Sedighi · M. Farsi
Cranfield School of Aerospace Transport & Manufacturing, Cranfield MK43 0AL, UK
e-mail: t.sedighi@cranfield.ac.uk

M. Farsi
e-mail: maryam.farsi@cranfield.ac.uk

© Springer International Publishing AG 2017
A. Hosseinian-Far et al. (eds.), *Strategic Engineering for Cloud Computing
and Big Data Analytics*, DOI 10.1007/978-3-319-52491-7_4

method with the method proposed in [2] using the 'roulette method'. We quantify the uncertainty of their density function, given the fact that the expert has only specified a limited number of probability judgements, and that these judgements are forced to be rounded. We will investigate the advantages of these methods against each other. Finally, we employ the proposed methods in this paper to examine the uncertainty about the prior density functions of the power law model's parameters elicited based on the imprecise judgements and how this uncertainty might affect our final inference.

1 Introduction

Assume the elicitation of a single expert's belief about some unknown continuous quantity denoted by θ. Identifying the underlying density function for θ based on the expert's beliefs, is the objective of the elicitation process. We denote this density function by $f(\theta)$. This function can be a member of a convenient parametric family. O'Hagan et al. [3] present a comprehensive review of the methods to elicit the density function chosen from a wide range of parametric families. Oakley and O'Hagan [1] criticised this way of eliciting expert's density function, and reported the following deficiencies as the main drawbacks of the so-called parametric elicitation methods: (1) The expert's beliefs are forced to fit the parametric family; (2) other possible distributions that might have fitted the elicited statements equally well have not been taken into account.

In order to tackle these drawbacks, they propose a nonparametric elicitation approach. In this approach, two parties are involved: an expert (female) and the analyst (male). The analyst receives the expert's elicited statements and will then make inference about the expert's density function, $f(\theta)$ based on her statements. In this approach, the analyst's fitted estimate of the expert's density function $f(\theta)$ is nonparametric, thereby avoiding forcing the expert's density into a parametric family.

This nonparametric elicitation approach can be seen as an exercise in Bayesian inference where the unknown quantity is $f(\theta)$. Moreover, the analyst's wishes are to formulate his prior beliefs about this unknown quantity. He then updates this prior in light of the data received from the expert in the form of probabilities assessments (e.g., quantiles, quartiles, mode, or mean) in order to obtain his posterior distribution for $f(\theta)$. The analyst's posterior mean serves as the 'best estimate' for the expert's density function $f(\theta)$, while the variance around this estimate describes the uncertainty in the elicitation process.

One crucial question that might arise in this non-parametric elicitation method is that, can the expert specify the probabilities assessments with absolute precision? This is evident that the facilitator cannot elicit her probabilities with absolute precision (or she cannot present her beliefs with the absolute precision). We are then interested to investigate what implications this might have on the uncertainty of the elicited prior distribution. There are several factors [3] that caused this issue. One of the most important challenges is that it is difficult for any expert to give precise numerical values for their probabilities; nevertheless, this is a requirement in most elicitation schemes. For instance, in many elicitation schemes, it is common to ask for percentiles or quantiles rather than probabilities. However, if the expert is unable to make precise probability assessments, then she will not be able to make precise percentile (or quantile) assessments either (see Sect. 3 and [3] for the differences between the 'precise' and 'imprecise' probabilities).

An obvious and natural source of imprecision comes from rounding done by the expert during probability assessments. We outline such imprecisions by the following explanation. Assuming expert's degree of belief probability $\theta < 0.1$, she may only consider probabilities rounded to the nearest, for instance, 0.05 (unless she wishes to state a probability close to 0 or 1). There may be an element of 'vagueness' in her imprecisely stated probabilities. She may believe that the event $\{\theta < 0.1\}$ is 'quite unlikely', yet she may find it difficult to attach a single numerical value to this belief. For example, she may find the question, "Is your probability of $\{\theta < 0.1\}$ less than 0.45?" straightforward to answer, but have difficulty in answering the question, "Is your probability of $\{\theta < 0.1\}$ closer to 0.1 or 0.2?", as she may have difficulty in deciding whether a probability of 0.1 describes her belief better than a probability of 0.2. The actual probability she reports may be chosen somewhat arbitrarily from the lower end of the probability scale, and all the subsequent probability judgements regarding θ will be dependent on this choice.

In order to investigate the impact of the imprecise probabilities stated by the expert on the elicited prior distribution and uncertainty around it (due to the imprecision), the elicitation methods must be modified in a way to capture this imprecision.

The purpose of this paper is to modify a *nonparametric* elicitation approach that is originally proposed by Oakley and O'Hagan [1] when the stated probabilities by the experts are imprecise. In Sect. 2, we briefly introduce this elicitation method and its properties and flexibility in practice. Section 3 is devoted to the short literature review of some main papers regarding the modelling of the imprecise probabilities. We also introduce a simple way for modelling the imprecision which is more straightforward to combine with the nonparametric elicitation method with a view to report the expert's density function and uncertainty around it. Section 4 is dedicated to the modification of Oakley and O'Hagan's elicitation method where the stated probabilities are not precise. Instead of eliciting percentiles or quantiles, we prefer to use a simpler type of fixed interval elicitation method known as *trial roulette* which is originally introduced in [4]. This method is introduced in Sect. 5. Finally, in Sect. 6, we employ the proposed methods in this paper to examine the uncertainty about the prior density functions of the power law model's parameters elicited based on the imprecise judgements and how this uncertainty might affect our final inference.

2   Nonparametric Elicitation

Let us consider eliciting an expert's beliefs about some unknown continuous variable θ. The elicitation is the process of translating someone's beliefs about some uncertain quantities into a probability distribution. The expert's judgements about the uncertain quantities are traditionally fitted to some member of a convenient parametric family. There are few deficiencies in this approach that can be tackled by the nonparametric approach proposed in [1], which will be briefly introduced in this section. In this approach, two parties are involved: expert and facilitator. We suppose that the elicitation is conducted by a (male) facilitator, who interviews the (female) expert and identifies a density function f that represents her beliefs about θ. He will help her as much as possible, for example by providing suitable training and discussing the various biases that can influence probability judgements.

The expert is usually only able to provide certain summaries of her distribution such as the mean or various percentiles. O'Hagan et al. [3] show that these information are not enough to determine $f(\theta)$, uniquely. In the traditional methods, a density function that fits those summaries, as closely as possible, is chosen from some convenient parametric family of distributions. Oakley and O'Hagan [1] reported two deficiencies regarding this approach: first, it forces the expert's beliefs to fit the parametric family; and second, it fails to acknowledge the fact that many other densities might have well fitted the same summaries equally. They then presented an approach that addresses both deficiencies together in a single framework.

This nonparametric approach allows expert's distribution to take any continuous form. The issue of fitting a density function to the summaries stated by the expert is considered as an exercise in Bayesian inference, where an unknown quantity of interest is the expert density function, denoted by $f(\theta)$. The facilitator formulates his prior beliefs about this quantity. He then updates his beliefs about the expert's true density function in light of the data that are obtained from the expert in the form of (probability) judgements. This is to obtain his posterior distribution for $f(\theta)$. The facilitator's posterior mean can then be offered as a 'best estimate' for $f(\theta)$, while his posterior distribution quantifies the remaining uncertainty around this estimate.

A very useful and flexible prior model for an unknown function (e.g., expert density function) is the Gaussian process. Thus, it is reasonable that the facilitator represents prior uncertainty about f using a Gaussian process: for any set $\{\theta_1, \ldots, \theta_n\}$ of values of θ, his prior distribution for $\{f(\theta_1), \ldots, f(\theta_n)\}$ is multivariate normal. The reason that the joint distribution of $f(.)$ at some finite points is multivariate normal comes from the formal definition of a Gaussian process over $f(\theta)$, and further details and examples can be seen in [3] and references therein. As $f(\theta)$ is a density function, two constraints are applied to the facilitator's prior: $\int_{-\infty}^{\infty} f(\theta)d\theta = 1$ and $f(\theta) \geq 0$ for all θ. The first constraint is applied as part of the data from the expert, and second constraint is applied (approximately) using simulation, which we discuss in later sections.

The facilitator specifies his mean and variance-covariance functions for f hierarchically, in terms of a vector α of hyperparameters. His prior expectation of $f(\theta)$ is some member $g(\theta \mid u)$ of a suitable parametric family with parameters u, contained within α, so

$$E(f(\theta) \mid \alpha) = g(\theta \mid u), \tag{1}$$

where $g(. \mid u)$ denotes the underlying density, in particular case, where $u = (m, v)$, is distributed as a normal density with mean m and variance v (t distribution can be considered as an alternative as suggested in [1, 5], but we do not consider this extension here), and

$$Cov\{f(\theta), f(\phi) \mid \alpha\} = \sigma^2 g(\theta \mid u) g(\phi \mid \phi) c(\theta, \phi \mid u, b), \qquad (2)$$

where $c(., . \mid u, b)$ is a correlation function that takes value 1 at $\theta = \phi$ and is a decreasing function of $|\theta - \phi|$, σ^2 specifies how close the true density function will be to its estimation, and so governs how well it approximates to the parametric family, and $\alpha = (m, v, b, \sigma^2)$. The following correlation function is used:

$$c(\theta, \phi \mid u, b) = \exp\{-\frac{1}{2vb}(\theta - \phi)^2\}, \qquad (3)$$

where b is the smoothness parameter. The correlation function above makes $f(.)$ infinitely differentiable with probability 1 (see [5] for further details and exploitation of this property).

The prior distribution of $\alpha = (m, v, \sigma^2, b^*)$ has the following form

$$p(m, v, \sigma^2, b^*) \propto p(m, v) p(\sigma^2) p(b^*). \qquad (4)$$

Oakley and O'Hagan [1] used an improper prior distribution for the Gaussian process variance σ^2. However, this prior makes the computation more feasible, but the analyst's posterior distribution of the expert density might be sensitive to the choice of prior of σ^2 (and b^* as well). We use the following proper prior distributions for σ^2 and b^*, respectively.

$$p(\sigma^2) = IG(a, d), \quad \log b^* \sim N(0, 1) \quad \text{or} \quad N(0, 4), \qquad (5)$$

where IG stands for Inverse Gamma.

We will also investigate sensitivity of the expert density with respect to changes of the prior distributions for σ^2 and b^*. For instance, other reasonable distributions for b^*, used in [1], are $N(0, 1)$ or $N(0, 4)$. The result of this study will be presented in the next section.

To update the facilitator's prior beliefs about the expert's density, we need to elicit some summaries as the data from the expert. O'Hagan et al. [3] recommended that the expert should be asked about the probabilities, percentiles or quantiles. However, the nonparametric approach is being used in this paper can be applied to any of probabilities, percentiles or quantiles, we generate the data by an elicitation scheme called 'trial roulette' that will be introduced in the next section. This method is quite simple to implement by the expert and originally suggested by Gore [4] in a medical case study. Suppose that the vector of data, D, is

$$D^T = \left(\int_{a_0}^{a_1} f(\theta)d\theta, \dots, \int_{a_{n-1}}^{a_n} f(\theta)d\theta \right), \tag{6}$$

where the end points $a_0, \dots, a_n$ denote the possible value of θ, and $a_i > a_{i-1}, i = 1, \dots, n$. It should be mentioned that D will also include this information that $\int_{-\infty}^{\infty} f(\theta)d\theta = 1$.

Because $f(\theta)$ has a normal distribution for all values of θ, then any data point also has a normal distribution and the joint distribution of D and any finite set of points on the function $f(\theta)$ is multivariate normal. The mean and variance of D are given by

$$H^T = E(D \mid (m, v))^T = \left(\int_{a_0}^{a_1} g(\theta \mid (m, v))d\theta, \dots, \int_{a_{n-1}}^{a_n} g(\theta \mid (m, v))d\theta \right),$$

and

$$Cov(P_{A_i}, P_{A_j} \mid \alpha) = \sigma^2 \int_{A_j} \int_{A_i} g(\theta \mid (m, v))g(\phi \mid (m, v))c(\theta, \phi \mid b^*, v)d\theta d\phi = \sigma^2 A, \tag{7}$$

where $A_i = [a_{i-1}, a_i]$ and $A_j = [a_{j-1}, a_i]$ for all $i, j = 1, \dots, n$. The details of the computation of H and A can be found in [6].

It then follows immediately from properties of the multivariate normal distribution that the updated facilitator's beliefs about $f(\theta)$ given D and α also has a normal distribution with

$$E(f(\theta) \mid D, \alpha) = g(\theta \mid m, v) + t(\theta \mid \alpha)^T A^{-1}(D - H), \tag{8}$$

and

$$Cov(f(\theta), f(\phi) \mid D, \alpha) = \sigma^2 \{ g(\theta \mid m, v)g(\phi \mid m, v)c(\theta, \phi \mid b^*, v) - t(\theta \mid \alpha)^T A^{-1} t(\theta \mid \alpha) \}, \tag{9}$$

where $t(\theta \mid \alpha)$ is given by

$$t(\theta \mid \alpha)^T = (Cov(f(\theta), P_{A_1} \mid \alpha), \dots, Cov(f(\theta), P_{A_n} \mid \alpha)),$$

$A_i = [a_{i-1}, a_i]$, and $\alpha = (m, v, b^*, \sigma^2)$.

In fact, conditional on α and data, the analyst's posterior distribution of $f(\theta)$ is again a Gaussian process with Eqs. (8) and (9) giving its mean and covariance structure, respectively.

The posterior distribution of α is given by

$$p(m, v, \sigma^2, b^* \mid D) \propto v^{-1}\sigma^{-n}|A|^{-\frac{1}{2}} \exp\{-\frac{1}{2\sigma^2}(D - H)^T A^{-1}(D - H)\}p(b^*)p(\sigma^2) \tag{10}$$

where n denotes the number of elements in D.

The conditioning on σ^2 can be removed analytically to obtain

$$p(m, v, b^* \mid D) \propto v^{-1} |A|^{-\frac{1}{2}} (\hat{\sigma}^2)^{-\frac{n}{2}} p(b^*) \tag{11}$$

where

$$\hat{\sigma}^2 = \frac{(D-H)^T A^{-1} (D-H)}{n-2}.$$

The conditioning on the rest of the hyperparameters cannot be removed analytically. We then use MCMC to obtain a sample of values of $\{m, v, b^*\}$ from their joint posterior distribution. Given a set of values for the hyperparameters, a density function is sampled at finite points of θ, $\{\theta_1, \ldots, \theta_k\}$, from the Gaussian process model. Repeating this many times, a sample of functions from $f(.) \mid D$ is obtained and the negative-valued functions are then removed. The remaining function are used to report estimates and pointwise credible bounds for the expert's density, $f(\theta)$.

3 Imprecise Probabilities: Modelling and Modification Issues

In this section, we briefly review the statistical literature of modelling of the imprecise subjective probabilities. We then present our approach to model the imprecision which exists in the expert's elicited probabilities. Finally, we modify the nonparametric elicitation approach presented in the previous section.

Imprecise probability models are needed in many applications of probabilistic and statistical reasoning. They have been used in the numerous scenarios such as:

- when there is little information for evaluating a probability (see Walley [7]);
- in robust Bayesian inference, to model uncertainty about a prior distribution (see Berger [10]; Pericchi and Walley [12]);
- in frequentist studies of robustness, to allow imprecision in a statistical sampling model, e.g., data from a normal distribution may be contaminated by a few outliers or errors that come from a completely unknown distribution (see Huber [13]);
- to account for the ways in which people make decisions when they have indeterminate or ambiguous information (see Smithson [14]).

Imprecise probability is used as a generic term to cover all mathematical models which measure chance or uncertainty without sharp numerical probabilities. It includes both qualitative and imprecise or nonadditive quantitative models. Most probability judgements in everyday life are qualitative and involve terms such as "probably" and "more likely than", rather than numerical values. There is an extensive amount of literature on these types of imprecise probability model.

Our main attempt here is to represent the imprecise probability stated by an expert as precise as possible in numbers. The most relevant work has been reported by

Walley [7], who considered bounding a probability P with upper and lower probabilities, $\overline{P}$ and $\underline{P}$ respectively. Unfortunately, this still leaves the issue of how to specify $\overline{P}$ and $\underline{P}$ with absolute precision unresolved. Additionally, the expert may also feel that values in the centre of the interval $[\underline{P}, \overline{P}]$ represent their uncertainty more appropriately than values towards the ends of the interval.

We express the stated probability $P^*(\theta \in A)$ as the expert's true probability plus an additive error, which represents the imprecision in the stated probability as follows,

$$P^*(\theta \in A) = P(\theta \in A) + \varepsilon, \tag{12}$$

where $\varepsilon \sim N(0, \tau^2)$ for some appropriate values of τ^2.

It should be noticed that normality itself can be considered as a strong assumption, but we now no longer have absolute lower and upper limits for the true probability $P(\theta \in A)$. This is also more plausible for the facilitator to give greater probability to value $P(\theta \in A)$ closer to $P^*(\theta \in A)$. The variance parameter τ^2 now describes the imprecision in the probability assessment, and may vary for different probability judgements. These values could be appropriately chosen in consultation with the expert.

Now, suppose that the expert provides k probability statements, denoted by $p_1^*, \ldots, p_k^*$ that are imprecise and modelled as follows:

$$p_i^* = p_i + \varepsilon_i, \quad i = 1, \ldots, k, \tag{13}$$

where p_i's are the expert's true probabilities that cannot be precisely stated, and ε_i is considered as a noise which has a distribution centred around zero and variance that describes the imprecision in the probability assessment.

The noise components in (13) can be followed of either normal distribution or uniform distribution. Oakley and O'Hagan [1] have briefly discussed the effect of including this noise with a specific variance for each noise on the elicited expert's density function.

In this section, the focus is to study the impact of a uniform noise on the expert's density function. The uniform noise is considered as $\varepsilon \sim U(-a, a)$, where the length of the interval $2a$, denotes the maximum possible error that would exist in the expert's stated probabilities.

We consider a type of fixed interval method known as trial roulette elicitation method originally proposed by Gore [4]. In this method, the expert is given n gaming chips and asked to distribute them amongst k bins. The proportion of chips allocated to a particular bin is representing her probability of θ lying in that bin, though clearly this probability is then subject to rounding error. This approach can be attractive to some experts, as they do not need to determine the probability statements directly, and the distributions of chips distributed between bins gives a histogram of their beliefs.

One element of this method is the lack of precision in the expert's stated probabilities. Suppose the expert distributes n given chips into k bins, then her stated probabilities must be multipliers of $1/n$. If the expert selects smaller n, then she is only able to make coarse probability assessments, so the precision will be lowered. Another issue to be considered is to locate the bins (particularly close to the end-points). The choice of scale can have a substantial effect on an expert's judgements [8] which is not further studied in this chapter.

We now modify the nonparametric elicitation method described above, by elicit-ing the imprecise probabilities using the trial roulette scheme. The k bins considered in this method are represented by the following intervals:
$[a_0, a_1), [a_1, a_2), \ldots, [a_{k-1}, a_k)$ (with the possibility of $a_0 = -\infty$ and $a_k = \infty$). The expert allocates n_i chips to the ith bin, $[a_{i-1}, a_i)$, with $\sum_{i=1}^{k} n_i = n$. We define $p_i^* = n_i/n$ to be the expert's stated probability of $\theta \in [a_{i-1}, a_i)$, with $p^* = (p_1^*, \ldots, p_k^*)$. These stated probabilities are usually subject to rounding errors which can be linked to the true probabilities through (13), where the noise components can be represented as $\varepsilon_i \sim U(-a, a)$, such that $\sum_{i=1}^{k} \varepsilon_i = 0$.

The facilitator's data consists of p^*, together with this knowledge that $\sum_{i=1}^{k} \varepsilon_i = 0$, and $f(\theta) \geq 0 \ \forall \theta$. The latter constraint implies $p_i \geq 0 \ \forall i$. We are now required to com-pute the facilitator's posterior distribution denoted by considering these conditions: $(\sum_{i=1}^{k} \varepsilon_i = 0, f \geq 0)$.

If the expert was to provide her true probabilities p rather than her stated probabili-ties p^*, it would be straightforward to derive the distribution of $f|p, \alpha = (m, v, b^*), \sigma^2$. This was discussed and presented in Sect. 2. However, it is more challenging to derive the posterior distribution of $f(\theta)|p^*, \alpha = (m, v, b^*, \sigma^2), \sum_{i=1}^{k} \varepsilon_i = 0)$, as H and A can-not be computed as presented above. In this scenario, we consider $\varepsilon = (\varepsilon_1, \ldots, \varepsilon_k)$ as extra hyperparameters. Therefore, the following prior distribution for the new set of hyperparameters (α, ε) would be considered as follows:

$$p(\alpha, \varepsilon \mid \Sigma_{i=1}^{k}\varepsilon_i = 0) \propto v^{-1} p(\sigma^2) p(b^*) p(\varepsilon \mid \Sigma_{i=1}^{k}\varepsilon_i = 0), \tag{14}$$

where $p(\sigma^2)$, $p(b^*)$ are the same priors presented in (5).

The main issue is to generate k random variables from $U(-a, a)$ such that $\Sigma_{i=1}^{k}\varepsilon_i = 0$. We propose the following procedure to generate the noise components:

- Let $\xi = 1 + \frac{k-2}{k}$;
- Draw a random sample of size k from $y_i \sim U(0, 1)$ and define $x_i = 2ay_i - a$. These randomly variables follow $U(-a, a)$;
- Let $\varepsilon_i = \frac{x_i - \bar{x}}{\xi}$, where $\bar{x} = \frac{1}{k}\Sigma_{i=1}^{k}x_i$.

The posterior distribution of (α, ε), after integrating out σ^2, is then given by

$$p(m, v, b^*, \varepsilon \mid D) \propto v^{-1} |A|^{-\frac{1}{2}} (\hat{\sigma}^2)^{-\frac{n+d}{2}} p(b^*) p(\varepsilon) \tag{15}$$

where $\hat{\sigma}^2 = \frac{1}{n+d-2}[a + (D + \varepsilon - H)^T A^{-1}(D + \varepsilon - H)]$.

To derive the expert's density function, we are similarly required to use MCMC to obtain samples of the hyperparameters, $\alpha = (m, v, b^*)$ from their joint posterior distribution. Weak priors are assumed for m and v as in (4). The following proposal distributions are chosen for the Metropolis–Hasting sampler:

$$m_t \mid m_{t-1} \sim N(m_{t-1}, 0.01)$$

$$\log(v_t) \mid m_{t-1}, m_t, v_{t-1} \sim N(\log(v_{t-1}), 0.1 \times (1 + 5 \times |m_t - m_{t-1}|))$$

$$\log(b_t^*) \mid b_{t-1}^* \sim N(\log(b_{t-1}^*), 0.01).$$

The chain is run for 20,000 iterations and the first 10,000 runs are discarded to allow for the burn-in period. A random density function is generated for each of the last 10,000 runs. Given a set of values for the hyperparameters, a density function is sampled at k points of θ, $\{\theta_1, \ldots, \theta_k\}$, from the Gaussian process model. Repeating this will give us a sample of functions from the posterior $p\{f(.) \mid D\}$, and estimated and pointwise credible bounds for $f(\theta)$ can then be reported.

4 A Gaussian Process Model for Imprecise Probabilities

In the trial roulette elicitation scheme described above, we only addressed the possible imprecision due to rounding errors. The more fundamental problem is that the expert does normally have the challenge of representing a feeling of uncertainty with a numerical value. The expert usually first states a small number of probability assessments about θ. She may then wish to draw her cumulative distribution or density function between these points. However, this would completely specify her distribution for θ over the chosen range, but it does not seem reasonable that the facilitator should know her distribution in this range with no uncertainty. In particular, if the expert's original probabilities were imprecisely given, she might then believe that other probabilities would represent her beliefs equally well, which would have led to different curves being drawn.

This motivates the development of a model for imprecision in which the facilitator will still have some uncertainty about the expert's true density function regardless of how many probabilities the expert provides. Therefore, there is a need to consider correlation in the expert's imprecision. In the case that the stated probability departs from her true probability by some amount at a particular value of θ, θ_0 say, a similar deviation at values of θ close to θ_0 would be then expected. In this section, we present an alternative way to model the imprecise probabilities, in which the error terms have a different distribution and correlation structure.

Suppose that $f^*(\theta)$ illustrates the expert's reported imprecise density function. The relationship between the $f^*(\theta)$ and her true distribution $f(\theta)$ can be given by

$$f^*(\theta) = f(\theta) + q(\theta), \tag{16}$$

where $q(\theta)$ is an *error* indicating the imprecision in the expert's probabilities; $q(\theta)$ is a distortion of the expert's true, precise probabilities that result in the imprecise reported probabilities.

Suppose that, the stated probabilities reported by the expert are presented as follows:

$$\mathcal{D} = (P^*_{B_1} = \int_{B_1} f^*(\theta)d\theta, \dots, P^*_{B_n} = \int_{B_n} f^*(\theta)d\theta, \int_{\Theta} q(\theta)d\theta = 0), \qquad (17)$$

where $P^*_{B_i}$ is the expert's stated probability for the ith interval, denoted by B_i, and $\Theta = \bigcup_{i=1}^{n} B_i$.

We believe both the expert's true distribution and reported distribution to be smooth functions, and we also specify a Gaussian process distribution of $q(\theta)$ that ensures that $q(\theta)$ will also be smooth. The variance of $q(\theta)$ is constructed in such a way that it will decrease when $f(\theta)$ approaches 0 or 1, so that the absolute magnitude of any imprecision error will decrease appropriately. Additionally, we condition on this assumption that the expert's imprecise probabilities still sum to 1. Therefore, the error term can be modelled by a Gaussian process with zero mean and the following covariance matrix,

$$Cov(q(\theta), q(\phi)) = \sigma_2^2 g(\theta)g(\phi)c_2(\theta, \phi).$$

As shown above, the facilitator's prior distribution of $f(\theta)$ can be also expressed by a Gaussian process:

$$f(\theta) \sim GP(g(\theta), \sigma_1^2 g(\theta)g(\phi)c_1(\theta, \phi)),$$

where $g(\theta) = N(m, v)$.

It can be shown that $f^*(\theta) \mid \mathcal{D}, m, v, b_1^*, b_2^*, \sigma_1^2, \sigma_2^2$ has a normal distribution with

$$E(f^*(\theta) \mid \alpha, \mathcal{D}) = g(\theta) + t(\theta)^T A^{-1}\{D - H\}$$

$$Cov\{f^*(\theta), f^*(\phi) \mid \alpha, \mathcal{D}\} =$$

$$\sigma_1^2\{g(\theta)g(\phi)c(\theta, \phi) - t(\theta)^T A_1^{-1}t(\phi)\} + \sigma_2^2\{g(\theta)g(\phi)c(\theta, \phi) - t(\theta)^T A_2^{-1}t(\phi)\}$$

where, $A = \sigma^1 A_1 + \sigma_2 A_2$, $b_i^* = \frac{b_i}{v}$, and details of calculating H, A_1, A_2 and t are available in [6].

The posterior distribution of $\alpha = (m, v, b_1^*, b_2^*, \sigma_1^2, \sigma_2^2)$ is easily found from the multivariate normal likelihood for $\mathcal{D}$ given α:

$$p(\alpha \mid \mathcal{D}) \propto v^{-1}|A|^{-\frac{1}{2}} \times \exp\{-\frac{1}{2}(D - H)^T A^{-1}(D - H)\}p(\sigma_1^2)p(\sigma_2^2)p(b_1^*)p(b_2^*), \quad (18)$$

where $p(\sigma_i^2)$ and $p(b_i^*)$ are the same priors presented in (5).

5 Applications in Reliability Analysis

In this section, we are going to investigate how allowing imprecision in the expert's probability assessments could influence the expert's density function regardless of its skewness to the right or left through some applications in reliability analysis. We focus on two main prior distributions that are widely used in Bayesian analysis of failure data: Beta distribution and Gamma distribution. The former one has been commonly used as a prior distribution to study success/failure data, while the latter one is the most popular conjugate prior for failure count data and failure time data.

5.1 Example: Beta Prior Distribution

In order to conduct a reliability assessment for a single item or for a system, analysts might require to obtain success/failure data. Such data is comprised of details surrounding the component's or system's success or failure while performing with a view to complete its intended function. For instance, testers may try an emergency diesel generator to see if it will start on demand, and record whether this generator starts or not. The record of whether the launched missile system completes its mission successfully or not is another example of such a dataset. This data can be modelled using binomial distribution for a fixed number of tests. For this model, the unknown parameter is the success probability, denoted by θ that must be estimated based on the test data and experts' knowledge.

The conjugate prior distribution for success/failure data is the beta distribution, denoted by $\mathscr{B}(\alpha, \beta)$ and is defined as follows:

$$f(\theta) = \frac{\Gamma(\alpha + \beta)}{\Gamma(\alpha) + \Gamma(\beta)} \theta^{\alpha-1}(1 - \beta)^{\beta-1}, \quad \theta \in [0, 1], \quad \alpha > 0, \quad \beta > 0,$$

where α is usually interpreted as the prior number of successful component tests and β as the prior number of failed component tests; and $\Gamma(.)$ stands for the Gamma function.

In this example, we focus on eliciting prior distribution for θ using the nonparametric prior elicitation described above and modelling the imprecision when the expert is asked to present her probability statements with trial roulette elicitation scheme. Without loss of generality, we suppose that the expert's true distribution is $\mathscr{B}(4, 2)$.

The trail roulette scheme is chosen to elicit the expert's probabilities. The method is explained to the expert and she is then asked to choose the number of bins and chips to express her belief. Assuming that the expert has allocated her chips precisely, as shown in Table 1, the facilitator has two sources of uncertainty about the expert's distribution. First, the expert has only provided four probabilities, corresponding to the four bins. This is illustrated in Figs. 1 and 2 which show the facilitator's pointwise 95% intervals for the expert's density and distribution functions (based on a

Table 1 Stated expert's probabilities given 20 chips and 4 bins. True probabilities are generated from a $\mathscr{B}(4,2)$ distribution

Num. of chips	$P(\theta \leq 0.4)$	$P(0.4 \leq \theta \leq 0.6)$	$P(0.6 \leq \theta \leq 0.8)$	$P(\theta \geq 0.8)$
10	0.10	0.20	0.40	0.30
20	0.10	0.25	0.40	0.25

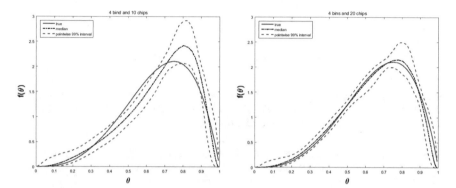

Fig. 1 The mean (*dashed-point line*) and pointwise 95% intervals (*dashed lines*) for the expert's density, and the true density (*solid line*) with uniform noise when the imprecise probabilities are included and rounded to nearest 5% (10 chips) and 2.5% (20 chips)

sample of 4000 functions), respectively. This issue is also investigated in Oakley [9] in details. The second source of uncertainty, which is of main interest in this paper, is due to the rounding in the expert's judgements. The effect of accounting for this as discussed above is also shown in Figs. 1 and 2, where the dashed lines show the facilitator's pointwise 95% intervals for the expert's density/distribution function, the solid lines show the expert true density/distribution function, and dot-dashed $(.-.-)$ line illustrate the expert's density/distribution function estimation based on the stated probabilities. Accounting for this imprecision has resulted in a wider range of distributions that might be consistent with the expert's beliefs.

The facilitator's uncertainty about the expert's PDF (probability density function) (or CDF (cumulative distribution function) when she uses 4 bins and 20 chips (i.e., the imprecisely stated probabilities are rounded to nearest 2.5%) to express her beliefs is smaller than the uncertainty of the corresponding elicited PDF (CDF), calculated based on 10 chips (where the probabilities stated by the expert are rounded to nearest 5%). Therefore, a combination of a small number of bins and large number of chips can result in fairly little uncertainty about the expert's density. This is to be expected: the Gaussian process model for f implies that the CDF is also a smooth function, so observing a small number of points on the CDF should reduce uncertainty substantially. In addition, it is trivial to conclude that including the noise in the data would affect the expert's density (see Oakley and O'Hagan [1] for another example and relevant details).

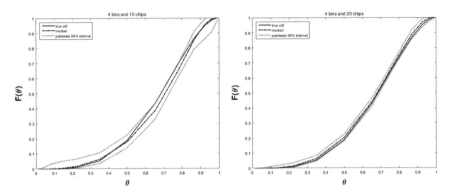

Fig. 2 The mean (*dashed-point line*) and pointwise 95% intervals (*dashed lines*) for the expert's CDF, and the true CDF (*solid line*) with uniform noise when the imprecise probabilities are included and rounded to nearest 5% (10 chips) and 2.5% (20 chips)

Table 2 The posterior mean, variance and their corresponding 95% pointwise bounds of the expert's density function using the trial roulette elicitation scheme with different number of chips

n	E_L	E_m	E_U	V_L	V_m	V_U	$E_U - E_L$	$V_U - V_L$
10	0.6938	0.7142	0.7335	0.0245	0.0310	0.0407	0.0397	0.0153
20	06836	0.6951	0.7062	0.0258	0.0302	0.0354	0.0226	0.0096
40	0.7047	0.7111	0.7173	0.0246	0.0267	0.0292	0.0126	0.0046

In other words, the approach proposed in this paper to modelling the imprecision would help the facilitator to measure the level of his uncertainty about the expert's density by decreasing/increasing the number of chips and/or bins. The elicitation task might become easier for the expert by choosing the suitable number of chips and bins. In the following table, we have shown that if 20 chips (instead of 10 chips) are used by the expert, then the uncertainty of the expert's density will be decreased by more than 50%. This rate will be more than 300%, if we use 40 chips against 10 chips. The corresponding results are given in Table 2. In this table, n stands for the number of chips used by the expert, and E_m and V_m are the mean and variance of the expert density function obtained by fitting a Gaussian process to her stated beliefs using the trial roulette scheme, respectively. In addition, E_L and E_U are the endpoints of 95% bound around the posterior mean of the fitted Gaussian process to the experts probabilities, and V_L and V_U are the endpoints of 95% bounds around the corresponding posterior variance.

5.2 Example: Gamma Prior Distribution

A commonly used prior distribution for the mean number of failures per unit time, and for the scale and shape parameters of the fitted distributions (e.g., exponential,

Table 3 Stated probabilities given the combinations of chips and bins. True probabilities are generated from a G(5,1) distribution

Num. of chips	$P_{(0,2)}$	$P_{(2,4)}$	$P_{(4,6)}$	$P_{(6,\infty)}$
10	0.10	0.30	0.30	0.30
20	0.05	0.30	0.35	0.30

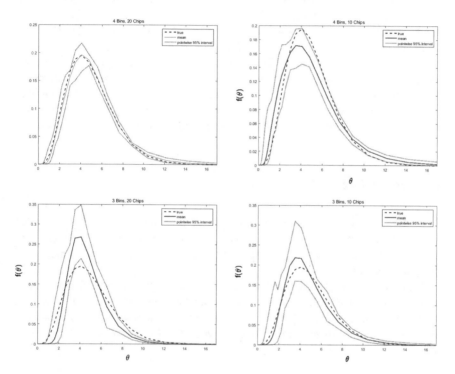

Fig. 3 Pointwise 95% intervals for the expert's density function (*dotted lines*), and the G(5,1) density function (*dashed lines*). *Solid lines* show the expert's estimated density using the nonparametric method

Gamma, Weibull) to model failure time data (the most commonly used data to assess component reliability) is the gamma distribution, which is denoted by $\mathscr{G}(\alpha, \beta)$ and presented as

$$f(\theta \mid \alpha, \beta) = \frac{1}{\beta^{\alpha}\Gamma\alpha}\theta^{\alpha-1}e^{-\frac{\theta}{\beta}}, \quad \theta > 0, \quad \alpha > 0, \quad \beta > 0.$$

Similar to the beta distribution example, we are interested in eliciting prior distribution for θ using the above elicitation method and modelling the imprecision, when the expert is asked to present her probability statements with trial roulette elicitation scheme. We assume the expert's true prior is $\mathscr{G}(5, 1)$. The expert allocates 10 and

20 into 4 bins as shown in Table 3. The effect of rounding the expert's probability assessments for this prior is illustrated in Fig. 3, where the dotted lines show the facilitator's pointwise 95% intervals for the expert's distribution function. Similar to the beta distribution example, accounting for this imprecision has resulted in a wider range of distributions that might be consistent with the expert's beliefs.

We also study what such intervals would be developed, when the expert state her true probability statements using different number of bins and chips. From Fig. 3, it can be concluded that the facilitator's uncertainty about the expert's PDF will grow by decreasing the number of chips and bins. This is simply because the expert is forced to round her true probabilities to the nearest number using smaller number of chips. Additionally, when we provide small number of bins, she cannot freely decide to allocate the right number of chips in a large bin. In [9, 11] we explored a close link between the number of chips and bins (used by expert to express her beliefs) and the uncertainty induced by the imprecision of experts' probabilities assessments.

6   Conclusions and Future Work

Acknowledging the imprecision in the expert's probabilities increases the uncertainty around the expert's true density. We have considered two ways to model imprecise probabilities: a uniform errors model; a Gaussian process model, as presented in (13).

We have used a very straightforward approach to implement the elicitation scheme so called trial roulette. However, it is crucial to investigate whether or not the trial roulette in itself is a good method for elicitation, and how it might work for unbounded parameters, in particular, when the stated probabilities are very imprecise.

References

1. Oakley, J. E., & O'Hagan, A. (2007). Uncertainty in prior elicitations: A nonparametric approach. *Biometrika*, *94*, 427–441.
2. Oakley, J. E., Daneshkhah, A., & O'Hagan, A. (2012). Nonparametric Prior Elicitation using the Roulette, submitted to Bayesian Analysis.
3. O'Hagan, A., Buck, C. E., Daneshkhah, A., Eiser, J. R., Garthwaite, P. H., Jenkinson, D., Oakley, J. E., & Rakow, T. (2006). *Uncertain judgements: Eliciting experts' probabilities*. Chichester: Wiley.
4. Gore, S. M. (1987). Biostatistics and the medical research council. *Medical Research Council News*.
5. Gosling, J. P., Oakley, J. E., & O'Hagan, A. (2007). Nonparametric elicitation for heavy-tailed prior distributions. *Bayesian Analysis*, *2*, 693–718.
6. Daneshkhah, A., & Sedighi, T. (2016). Nonparametric prior elicitation with imprecisely assessed probabilities (Working Paper).
7. Walley, P. (1991). *Statistical reasoning with imprecise probabilities*. London: Chapman and Hall.

8. Garthwaite, P. H., Jenkinson, D. J., Rakow, T., & Wang, D. D. (2007). *Comparison of fixed and variable interval methods for eliciting subjective probability distributions*, Technical report, University of New South Wales.
9. Oakley, J. E. (2010). Eliciting univariate probability distributions. In K. Böcker (Ed.), *Rethinking risk measurement and reporting* (Vol. I). London: Risk Books.
10. Berger, J. O. (1990). Robust Bayesian analysis: Sensitivity to the prior. *Journal of Statistical Planning and Inference, 25*, 303–328.
11. Daneshkhah, A., & Oakley, J. E. (2010). Eliciting multivariate probability distributions. In K. Böcker (Ed.), *Rethinking risk measurement and reporting* (Vol. I). London: Risk Books.
12. Pericchi, L. P. and Walley, P. (1991). Robust Bayesian credible intervals and prior ignorance. *International Statistical Review, 58*, 1–23.
13. Huber, P. (1981). *Robust Statistics*, New York: Wiley.
14. Smithson, M. (1989). *Ignorance and Uncertainty: Emerging Paradigms*, New York: Springer-Verlag.

Part II
Systemic Modelling, Analysis and Design for Cloud Computing and Big Data Analytics

Early Detection of Software Reliability: A Design Analysis

R. Selvarani and R. Bharathi

Abstract Reliability of software is the capability of itself to maintain its level of stability under specified conditions for a specified period of time. The reliability of software is influenced by process and product factors. Among them, the design mechanism has a considerable impact on overall quality of the software. A well-designed internal structure of software is a required for ensuring better reliable. Based on this, we propose a framework for modeling the influence of design metrics on one of the external quality factors, reliability of the software. Here, multivariate regression analysis is applied to arrive a formal model, which is the linear combination of weighted polynomial equations. These estimation equations are formed based on the intricate relationship between the design properties of software system as represented by CK metric suite and the reliability.

Keywords Object-oriented · Reliability · Design metrics · Influence

1 Introduction

Reliability is one of the important characteristics of software, which can be measured quantitatively by analyzing the failure data and it is defined as "The probability of failure free operation of a computer program in a specified environment for a specified period of time" [1]. Software reliability is generally accepted as one of the major factors in software quality since it quantifies software failures, which can make a powerful and critical system inoperative [2]. Mostly, the software failures are due to poor design quality. A significant research effort has been spared to define the reliability at design phase of SDLC. The goal of this work is to develop

R. Selvarani
Computer Science and Engineering, Alliance University, Bangalore, India
e-mail: selvarani.r@alliance.edu.in

R. Bharathi (✉)
Information Science and Engineering, PESIT-BSC, Bangalore, India
e-mail: rbharathi@pes.edu

© Springer International Publishing AG 2017
A. Hosseinian-Far et al. (eds.), *Strategic Engineering for Cloud Computing and Big Data Analytics*, DOI 10.1007/978-3-319-52491-7_5

an estimation technique to estimate the reliability of object-oriented software modules at design stage of SDLC and provides a feedback to the design architect to improve the design. The most accepted and frequently used design metric in object-oriented (OO) technology is CK metrics [3, 4, 5]. In this work, we have proposed a framework for estimating the reliability at the design phase of SDLC. A valid hypothesis is established relating the design metrics and reliability through empirical analysis. A functional relationship is established through multivariate regression technique for selected design metrics. With these equations, an estimation model called R-MODEL is formulated based on the intricate relationship existing between the selected set of design metrics (CK metrics) and the reliability. The method is developed in such way that the reliability estimation as an integral part of the software design phase.

Terminologies: IEEE standard [6] defines "Fault" is a defect, which can be the cause of one or more failures. A fault is an error that can be fixed during design stage. "Failure" is defined as the incapability of the system or system component to perform a required function within specified limits. "Error" is defined as the human negligence that results in a fault.

The chapter is organized as follows: Sect. 2 explains the background of this research work. Secttion 3 discusses on CK metric and the NASA recommendations. Section 4 describes the proposed framework for the estimation of software reliability at the design phase. Section 5 elaborately discusses the empirical study carried out for modeling the influence of design metrics on reliability. Section 6 explains the influence of design metrics on software reliability. Section 7 demonstrates the R-MODEL through design metrics. Section 8 illustrates the results of model evaluation and validation process, followed by threats to validity in Sects. 9 and 10 summarizes the conclusion.

2   Background

The software reliability prediction is an assessment of the software product based on parameters associated with the software and its development environment [6]. Mohanta et al. [7] developed a model based on Bayesian belief network. Using the identified design metrics they determined the reliability of a class and then the overall system reliability is predicted based on use case reliabilities and operational profile. Hong-Li Wu et al. [8] have proposed software reliability prediction model built using benchmark measurement. Steinberg et al. [9] cited that the reliability could be measured based on defect rate of the software.

A software metric based reliability models was proposed using static analysis in [10]. Roshandel et al. [11] proposed a framework for estimating the component reliability at the software architecture level. In this framework, initially they used the dynamic behavior model as a basis for estimation framework, and then they utilized view-level inconsistencies between the different architectural models to obtain architectural defects, since these defects are used to model the failure behavior of the component. Peng Xu et al. [12] have provided a reliability model for OO software considering software complexity and test effectiveness. They have also stated that software complexity factor can be calculated during the design phase of SDLC, thereby helping developers to design the software to meet the required level of reliability. Norman [13] has reported that as per American Institute of Aeronautics and the Institute of Electrical and Electronics Engineers, the reliability can be predicted over all phases of the software life cycle, since early error identification reduces the cost of error correction. Further the author mentioned that the premise of most reliability estimation models is based on failure rate that is a direct function of the number of faults in the program. An enhanced model to predict reliability at early stages has been proposed in [14]. To estimate the reliability, they have predicted the number of faults at various stages of development. In this work, we refer our proposed empirical model for estimating the reliability using weighted polynomial regression equations with appropriate mapping of design metrics toward reliability [15].

3 Design Metrics

The internal design metrics have a strong association with the quality of the software [16]. Many researchers [3, 4, 17–20] have clearly mentioned that CK metric suite is the good indicator of quality factors in OO software. With this motivation, the following selected set of CK metrics [21] are used to predict the reliability of the software at the design phase of SDLC.

Depth of Inheritance Tree (DIT): It is the length of the longest path from a given class to the root class in the inheritance hierarchy. The deeper a class is in the hierarchy, the greater the number of methods it is likely to inherit, making it more complex to predict its behavior.

Response for a class (RFC): It is a set of methods that can be potentially be executed in response to a message received by an object of a particular class. The larger the number of methods that can be invoked from a class, the greater the complexity of the class.

Number of Children (NOC): It is the number of immediate subclasses subordinated to a given class in the class hierarchy. The improper abstraction of the parent class is the result of higher number of children.

Weighted Methods per class (WMC): It is the sum of complexities of methods in a class.

Coupling between Object classes (CBO): It is a count of the number of classes to which it is coupled. The larger the number of couples, the higher the sensitivity to changes in other parts of the design.

The design stage of an OO software system is the vital and critical stage of SDLC, as it dictates the quality levels of developed software [22]. To get a better quality factor the classes have to be designed with appropriate levels of design

Table 1 NASA threshold for
CK metrics

CK metrics	NASA threshold
DIT	3, Maximum 6
RFC	(50, 100), Maximum 222
NOC	Maximum 5
CBO	5, Maximum 24
WMC	20, Maximum 100

complexity metrics. The careful designing of metrics will result in defect free methods and classes, which results in a highly reliable system [15].

It is also observed that any reliability prediction model depends on the direct function of the number of faults. Mitigating the faults can control the failure rate in the software. The occurrence of these faults is due to improper designing of OO software [22].

In this paper, the influence of the CK metric suite on reliability as a function of defect proneness is being analyzed. The threshold set recommended by NASA research laboratory [23] is considered for our analysis (Table 1).

In order to estimate the reliability with respect to the defect proneness of the design, an empirical analysis is carried out for mapping the internal design metrics with reliability.

4 Reliability Estimation Framework

Figure 1 is the proposed framework for reliability estimation model. Using the UML class diagram or the equivalent Java code, the CK metric data are obtained. Automated tools like Rational Rose, etc., can be used to get these metric data from code artifacts. Manual method of counting is adopted when the input for the measurement system is the class diagram. DIT, NOC and WMC can be calculated using static class diagrams and metrics like RFC and CBO can be calculated using dynamic sequential diagrams [22 and 15].

The empirical analysis is carried out on these metric data obtained from various medium high-level projects from service-based industries. Table 2 depicts the data related to empirical analysis. The influence of individual metrics on reliability is arrived through this analysis as shown in Figs. 2 and 3. The relationship between these selected metrics and reliability is mapped through multifunctional equations as shown in Eq. (2). The final estimation model called R-MODEL is arrived through the linear combination of these multifunctional estimation equations with appropriate weighted coefficients as shown in Eq. (3).

When the CK Metric data (captured from design data) of any software developed based on OO technology is given as input to this model, the reliability of the software can be estimated. This is a novel approach for estimating the reliability of the software at an early stage. This will provide a feedback on reliability for the

Fig. 1 Framework for
reliability estimation

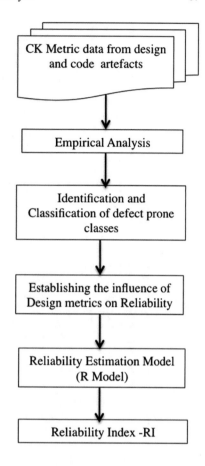

software architect at the design stage. Hence the design may be modified for the
required/expected level of reliability before implementation through code [15].

5   Empirical Model

An empirical research seeks to explore, describe, predict, and explain natural, social
or cognitive phenomena by using evidence based on observation or experience [24].
Kitchenham et al. [25] have proposed preliminary guidelines for empirical research
in software engineering. In our research, the mapping between two variables viz.,
reliability and design metrics is arrived through experimental data analysis. The
data presented in relation to the defect proneness of the software at the design stage
in the research work [22] is considered here for further analysis. The empirical
analysis is carried out is based on the following hypothesis,

H_R: The lower the value of design metrics the higher is the reliability.

Table 2 Empirical model

Metric	N	Project 1			Project 2			Mean
		Bin	NDC	Rel$_d$	Bin	NDC	Rel$_d$	Mean value
DIT	1	58	8	96.64	128	15	97.45	97.05
	2	68	14	94.12	150	26	95.59	94.85
	3	29	24	89.92	91	44	92.53	91.22
	4	38	50	78.99	82	95	83.87	81.43
	5	29	63	73.53	80	119	79.80	76.66
	6	16	71	70.17	59	72	87.78	78.97
RFC	10	68	2	99.16	89	3	99.49	99.33
	30	123	4	98.32	249	9	98.47	98.40
	55	27	8	96.64	98	15	97.45	97.05
	70	6	11	95.38	45	23	96.10	95.74
	85	5	14	94.12	48	26	95.59	94.85
	100	2	16	93.28	35	28	95.25	94.26
	140	3	37	84.45	22	75	87.27	85.86
	160	2	64	73.11	1	117	80.14	76.62
	200	2	70	70.59	2	132	77.59	74.09
NOC	0	228	0	100.0	330	0	100.0	100.00
	1	3	8	96.64	175	15	97.45	97.05
	2	2	20	91.60	56	37	93.72	92.66
	3	2	35	85.29	12	66	88.79	87.04
	4	2	59	75.21	5	102	82.68	78.95
	5	1	74	68.91	11	149	74.70	71.81
CBO	1	2	4	98.32	5	7	98.81	98.57
	2	3	5	97.90	4	10	98.30	98.10
	4	31	7	97.06	218	7	98.81	97.94
	5	20	11	95.38	20	13	97.79	96.59
	8	71	19	92.02	71	37	93.72	92.87
	10	30	36	84.87	109	68	88.46	86.66
	15	40	47	80.25	40	88	85.06	82.66
	20	30	61	74.37	112	114	80.65	77.51
	24	11	76	68.07	11	144	75.55	71.81
WMC	1	9	3	98.74	9	6	98.98	98.86
	5	72	5	97.90	84	11	98.13	98.02
	10	67	8	96.64	167	15	97.45	97.05
	20	60	12	94.96	183	22	96.26	95.61
	40	20	19	92.02	45	36	93.89	92.95
	60	7	30	87.39	58	59	89.98	88.69
	80	1	53	77.73	50	100	83.02	80.38
	100	1	70	70.59	1	132	77.59	74.09
	110	1	76	68.07	1	144	75.55	71.81

Total Number of Classes (ToC): 238 Total Number of Classes (ToC): 589
N: Metric Value
Bin: Bin of classes
NDC: Number of Classes contributed to defects from the bin having metric = N
Rel$_d$: Percentage influence on reliability

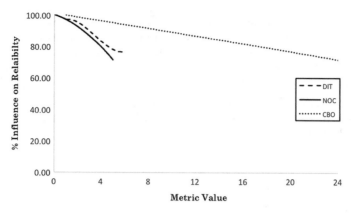

Fig. 2 Influence of DIT, NOC, CBO on reliability

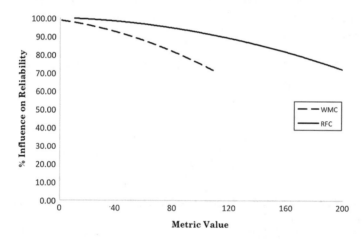

Fig. 3 Influence of WMC and RFC on reliability

Classes with low design metric values will be fewer defects prone and defect proneness is functionally related with reliability. The defect proneness of a class is defined as "the possibility of occurrence of fault" and it is estimated through heuristics and empirical analysis [17, 22].

Project I consists of 78 defective classes out of 238 classes and project II consists of 147 defective classes out of total 589 classes. These data are adopted from the work of Nair [22]. The details of the projects are not mentioned here as per the Non-Disclosure agreement. The classes in these two projects are segregated and grouped according to different CK metric values. It is assumed that the projects were designed taking into account the NASA threshold values for each metric. And these values are depicted in Table 2. Researchers [17, 22] proved that each CK metric has a range of values that manifest in the design of the software.

The static class diagram is used to calculate DIT, WMC and NOC metrics. Metrics like RFC and CBO are calculated using sequential diagrams. The bin of classes having metric value = N are manually investigated for its design properties. The bins are dependent on the nature and the behavior of the design metrics. Then the number of classes contributed to defects from the bin of classes having metric value = N (NDC) is calculated [15]. Finally the percentage reliability with respect to defect proneness (Rel$_d$) contributed by each metric is being calculated and depicted in Table 2.

5.1   Calculation of Rel$_d$

The percentage influence of design metrics on reliability (Rel$_d$) is quantitatively calculated based on the number of defect prone classes [15]. Hence the percentage influence on reliability becomes a function of defect prone classes.

Percentage influence on reliability

$$Rel_d = 1 - \text{Function (defect prone classes)}$$
$$Rel_d = 1 - f(DpC)$$

Rel$_d$ for each design metric:

$$Rel_{dm} = \left(1 - \left(\frac{DpC}{ToC}\right)\right) * 100, \quad \text{where m is metric} \tag{1}$$

DpC   Number of Defective Classes (Classes contributed to defects from the bin having metric = N)

ToC   Total number of classes.

For example in a project I, the total number of classes = 238, when DIT = 5, the number of classes contributed to defects from the bin is 63. Substituting these values in Eq. (1), we get Rel$_{dm1}$ = 73.5%, where m$_1$ is DIT.

Similarly, for each value of design metric the percentage influence on reliability is calculated.

6   Influence of Design Metrics on Reliability

It is observed that the influence of DIT on reliability is in the range of 91–78% when DIT varies from 3 to 6. It is clear that the DIT and the influence factor on reliability are inversely proportional. Hence it is hypothesized that the classes with high DIT value will have more defects and hence the reliability is reduced.

From the Table 2, it is observed that the reliability is high when NOC is less. When NOC is 1, reliability is 97%, and decreases to 78% when NOC is 4. The Table 2 shows the empirical results related to CBO on reliability. It is observed that when CBO is 15, the percentage influence on reliability is 82 and it becomes 77 when CBO equals to 20. It is evident that when CBO increases, the reliability of software decreases. The relationship curves for these three metrics are presented in Fig. 2.

It is observed from the Table 2, that when RFC was 100, the percentage influence on reliability is 94 and when RFC is increased to 120, it becomes 85. When RFC is high, the defect prone classes are more and hence the reliability will be low. Table 2 depicts that, reliability is high when WMC is less. When WMC is 20, the percentage influence on reliability is 95 and when WMC is increased to 40, reliability becomes 92. Figure 3 shows the relationship curve for these two metrics on reliability.

Hence it is hypothesized that low values of design metrics will result in high reliability.

7 R-MODEL Estimation

Based on the above discussions it is evident that there exist a functional relationship among the design quality metrics and software reliability. A successful model for the functional relationship between them is generated through polynomial regression analysis where the set of equations fit into the data distribution between them. Thereby, a significant relationship between the reliability (dependent variable) and design metrics (independent variables) is established.

The regression models are fitted using the method of least squares. Symbols α, β, γ, φ, Ψ in the functional estimation equations represent the independent variables such as DIT, RFC, NOC, CBO and WMC respectively. The functional estimation equations are presented in Eq. (2). R_α means Reliability due to DIT.

$$
\begin{aligned}
R_\alpha &= 0.4128\alpha^3 - 4.3814\alpha^2 + 8.4709\alpha + 92.496 \\
R_\beta &= -0.0006\beta^2 - 0.0178\beta + 99.936 \\
R_\gamma &= -0.6388\gamma^2 - 2.5456\gamma + 100.14 \\
R_\phi &= -0.0026\phi^2 - 1.15316\phi + 100.99 \\
R_\psi &= -0.0013\psi^2 - 0.1037\psi + 98.641
\end{aligned}
\tag{2}
$$

The above equations are best-fit functions for the set of selected design metrics. From the functional mapping between these two parameters it is observed that if the design complexity increases the reliability will be reduced because of the increase in defect prone classes. The above formal model Eq. (2) calculates the influence factor as shown in Table 3. R-DIT, R-RFC, R-NOC, R-CBO, R-WMC are the

Table 3 CK metric data and influence of design metrics on reliability

Module	Class	DIT	RFC	NOC	CBO	WMC	R-DIT	R-RFC	R-NOC	R-CBO	R-WMC	Class Rel
I	1	3	22	1	6	7	90.0717	99.254	96.9556	93.9778	97.8514	92.841827
	2	2	61	1	5	22	95.4146	96.6176	96.9556	95.1595	95.7304	93.118204
	3	3	6	0	9	2	90.0717	99.8076	100.14	90.4015	98.4284	93.061093
	4	1	6	0	5	7	97.0483	99.8076	100.14	95.1595	97.8514	95.127687
	5	4	58	0	2	25	83.4964	96.8852	100.14	98.6734	95.236	92.077768
	6	4	9	0	9	4	83.4964	99.7272	100.14	90.4015	98.2054	91.751106
	7	1	0	0	6	1	97.0483	99.936	100.14	93.9778	98.536	95.077581
	8	1	3	0	12	5	97.0483	99.8772	100.14	86.7784	98.09	93.680729
	9	1	4	0	10	6	97.0483	99.8552	100.14	89.199	97.972	94.088437
	10	4	6	0	9	5	83.4964	99.8076	100.14	90.4015	98.09	91.744106
	11	2	12	0	19	4	95.4146	99.636	100.14	78.1425	98.2054	91.790704
	12	3	13	0	13	6	90.0717	99.6032	100.14	85.5603	97.972	92.057517
	13	3	16	0	19	12	90.0717	99.4976	100.14	78.1425	97.2094	90.548673
	14	1	3	0	12	3	97.0483	99.8772	100.14	86.7784	98.3182	93.726369
	15	3	5	0	19	7	90.0717	99.832	100.14	78.1425	97.8514	90.743953
	16	3	5	0	6	8	90.0717	99.832	100.14	93.9778	97.7282	93.569667
	17	3	23	0	15	10	90.0717	99.2092	100.14	83.1085	97.474	91.437793
	18	1	19	0	12	30	97.0483	99.3812	100.14	86.7784	94.36	92.835529
	19	4	31	0	16	22	83.4964	98.8076	100.14	81.8748	95.7304	89.53738
	20	1	7	0	3	2	97.0483	99.782	100.14	97.5073	98.4284	95.660571
	21	3	1	0	19	1	90.0717	99.9176	100.14	78.1425	98.536	90.897993
	22	3	20	0	11	10	90.0717	99.34	100.14	87.9913	97.474	92.342857
	23	4	23	0	15	10	83.4964	99.2092	100.14	83.1085	97.474	90.188486
	24	1	2	0	2	4	97.0483	99.898	100.14	98.6734	98.2054	95.849069
	25	3	3	0	12	2	90.0717	99.8772	100.14	86.7784	98.4284	92.422855

(continued)

Table 3 (continued)

Module	Class	DIT	RFC	NOC	CBO	WMC	R-DIT	R-RFC	R-NOC	R-CBO	R-WMC	Class Rel
Average												
II	1	1	2	0	11	1	91.4658	99.411	99.8852	88.3534	97.5754	92.656434
	2	1	3	0	14	16	97.0483	99.898	100.14	87.9913	98.536	93.992411
	3	2	4	0	9	7	97.0483	99.8772	100.14	84.337	96.649	92.953077
	4	1	21	0	7	14	95.4146	99.8552	100.14	90.4015	97.8514	93.970364
	5	3	26	0	24	10	97.0483	99.2976	100.14	92.7909	96.9344	94.415939
	6	2	1	0	22	1	90.0717	99.0676	100.14	71.818	97.474	89.377183
	7	1	25	0	16	9	95.4146	99.9176	100.14	74.3634	98.536	91.232906
	8	1	60	0	20	31	97.0483	99.116	100.14	81.8748	97.6024	92.548321
	9	1	11	0	25	4	97.0483	96.708	100.14	76.888	94.177	90.484017
	10	1	43	1	9	21	97.0483	99.6676	96.9556	90.4015	98.2054	90.738527
	11	2	12	0	12	3	95.4146	98.0612	100.14	86.7784	95.89	92.892807
	12	2	2	0	19	2	95.4146	99.636	100.14	78.1425	98.3182	93.367726
	13	2	10	0	15	7	95.4146	99.898	100.14	83.1085	98.4284	91.887704
	14	1	5	0	16	3	97.0483	99.698	100.14	81.8748	97.8514	92.626184
	15	3	43	0	22	15	90.0717	99.832	100.14	74.3634	98.3182	92.834681
	16	2	6	0	11	2	95.4146	98.0612	100.14	87.9913	96.793	89.497875
	17	1	48	0	10	28	97.0483	99.8076	100.14	89.199	98.4284	93.642408
	18	2	3	0	12	5	95.4146	97.6992	100.14	86.7784	94.7182	93.006477
	19	1	38	0	16	21	97.0483	99.8772	100.14	81.8748	98.09	93.370326
	20	3	8	0	18	3	90.0717	98.3932	100.14	79.3918	95.89	92.061281
								99.7552			98.3182	91.046827

(continued)

Table 3 (continued)

Module	Class	DIT	RFC	NOC	CBO	WMC	R-DIT	R-RFC	R-NOC	R-CBO	R-WMC	Class Rel
Average							95.4300	99.2062	99.9807	82.5453	97.35048	92.297352
III	1	6	31	0	19	6	76.5558	98.8076	100.14	78.1425	97.972	87.995172
	2	4	7	0	6	10	83.4964	99.782	100.14	93.9778	97.474	92.25952
	3	4	27	0	15	1	83.4964	99.018	100.14	83.1085	98.536	90.362646
	4	2	25	1	12	9	95.4146	99.116	96.9556	86.7784	97.6024	92.483686
	5	3	6	1	16	8	90.0717	99.8076	96.9556	81.8748	97.7282	90.749367
	6	1	34	1	3	3	97.0483	98.6372	96.9556	97.5073	98.3182	94.772691
	7	3	12	0	19	2	90.0717	99.636	100.14	78.1425	98.4284	90.820153
	8	2	4	0	11	15	95.4146	99.8552	100.14	87.9913	96.793	93.324848
	9	5	35	0	15	6	78.1655	98.578	100.14	83.1085	97.972	89.148975
	10	4	26	0	2	14	83.4964	99.0676	100.14	98.6734	96.9344	92.853928
	11	5	11	0	1	8	78.1655	99.6676	100.14	99.8343	97.7282	92.328779
Average							86.4906	99.2703	99.2715	88.1036	97.77153	91.554524

Table 4 Statistics of Input Variables- A summary

Project	No of modules	Variable	Mean	Std. dev.	Min	Max
I	25	DIT	2.48	1.16	1	4
		RFC	14	16.24	0	61
		NOC	0.08	0.2768	0	1
		CBO	10.64	5.415	2	19
		WMC	8.6	7.88	1	30
II	20	DIT	1.65	0.745	1	3
		RFC	18.55	18.409	1	60
		NOC	0.05	0.2236	0	1
		CBO	15.4	5.335	7	25
		WMC	10.15	9.178	1	31
III	11	DIT	3.55	1.54	1	6
		RFC	10.8	6.75	1	19
		NOC	19.82	11.91	4	35
		CBO	0.27	0.47	0	1
		WMC	7.455	4.525	1	15

reliability factor which are defined as the percentage influence of the individual design metrics on reliability.

Three different modules of a commercial Java projects are selected for this analysis. Because of NDA, the details of the industry and data are not provided in this paper. Table 3 depicts the associated design metrics in different modules, and simultaneously depicts the computed data for reliability and Table 4 gives a summary statistics of input variables.

From Table 3, it is observed that when DIT is low, R-DIT is high. The value ranges from 83.49 to 97.04. Similarly R-RFC varies from 96.88 to 99.91 for the RFC values 1 to 58 respectively. For NOC values 0 and 1, R-NOC varies from 96.95 to 100. The influence of CBO (R-CBO) is high when CBO is 2 and low when CBO is 19. The value of R-WMC is 95.07, when WMC is 1 and for 30 the value of R-WMC is 92.83. All these influences support our hypothesis H_R.

From these discussions it is observed that, on an average the influence of each metric on reliability as a function of defect proneness is almost equal and varying from 88.3 to 99.4 as shown in Table 3. Consequently we assume that the weighted contribution of each selected metric on reliability is approximately 0.2 on unit scale.

R-DIT, R-RFC, R-NOC, R-CBO, R-WMC- $\rightarrow$ Percentage influence of DIT, RFC, NOC, CBO and WMC respectively on Reliability.

Class-Rel—Class Reliability

The weighted contribution of each metric on unit scale is arrived through statistical analysis. The concerned design metrics are distributed with weights 0.2 for all metrics. These weighted parameters demonstrate the relative significance of each design metric on reliability. Based on this analysis, the reliability index (RI) is derived as,

RI = f (DIT, RFC, NOC, CBO, WMC)

   = f($\alpha, \beta, \gamma, \varphi, \Psi$)

$$RI = 0.19\left(\frac{1}{n}\right)\sum_{r=1}^{n} R_\alpha + 0.2\left(\frac{1}{n}\right)\sum_{r=1}^{n} R_\beta + 0.2\left(\frac{1}{n}\right)\sum_{r=1}^{n} R_\gamma + 0.2\left(\frac{1}{n}\right)\sum_{r=1}^{n} R_\phi + 0.2\left(\frac{1}{n}\right)\sum_{r=1}^{n} R_\psi$$

$$(3)$$

Class-Reliability (Class-Rel) is arrived through the R-MODEL and the results are depicted in Table 3. These are the RI of the selected projects. These estimated values of reliability will provide a feedback to the design architect for improvising the design before implementation.

8 Model Evaluation and Validation

For evaluating and validating our model three different commercial C++ and Java projects are selected. The Table 4 summarizes statistics of the input variables.

These three different modules are tested using software testing tools and the reliability is estimated at industry. This estimated reliability at industry is compared with our estimated reliability using R-MODEL as shown in Fig. 4.

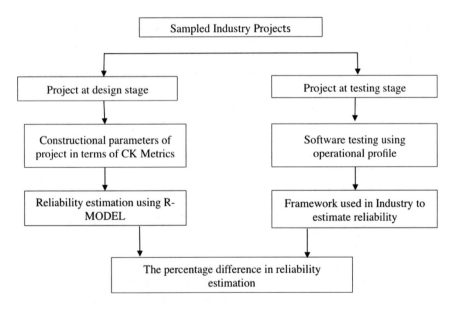

Fig. 4 Model evaluation and validation

Table 5 R-MODEL performance–a comparison

Modules	R-MODEL (1)	Industry data (2)	Difference between 1 and 2
1	92.6471	91.4237	1.2234
2	92.2974	90.9763	1.3211
3	91.5545	89.7589	1.7956

A comparison of R-MODEL with industry data is shown in Table 5.

9 Threats to Validity

9.1 Internal Validity

This study was conducted practically on large number of classes from various service projects, which are wide enough to support our findings. It is anticipated that these projects displayed the maximum probable reliability nature in these analysis. If the reliability assessment pattern appeared out of these classes is not sufficient enough or any noticeable fault that might usually repeat for unknown reasons in the modules, may call for a threat to validity in generalizing this method.

From the theoretical approach, it is observed that each metric has a defined relationship with reliability. The internal threat could be the interdependencies of each metric, however the studies conducted over the set of software tend to relieve this. It demonstrates a normal dependency. Therefore the threat to internal validity remains least.

In this study we engaged a team of engineers have similar experience and training. This avoids maturity level threat.

9.2 External Validity

There could be threat, that considered projects may not be fully object oriented. It can be avoided by careful inspection on design patterns. The considered projects are commercial projects only. If the projects are in a specialized area, then there will be a major effect, it is presumed. This threat requires dealing with projects of same application only.

10 Conclusions

In this chapter, we have discussed a data-driven approach for estimation of reliability index for the classes at the design phase of SDLC. This framework computes the reliability index taking into account, a set of selected design complexity metrics represented by the CK metrics, which in turn can be termed as the influence of the metric suit on reliability. Through empirical analysis, we have established the validity of hypotheses and subsequently we developed a formal model called R-MODEL through the multivariate regression analysis, which is capable of computing the reliability index of different modules of a project. This model was evaluated and found to be fairly matching with the industrial data. The early detection of this quality factor in a software system will effectively support in creating cost effective better quality software. This sort of analytical methods can be incorporated into futuristic autonomic design scenario.

References

1. Musa, J. D, Iannino, A., & Okumuto, K. (1987). Engineering and managing software with reliability measures. McGraw-Hill.
2. Lyu, M. R. (2007). Software Reliability Engineering: A Roadmap. *IEEE Computer Society, Future of Software Engineering* (FOSE'07).
3. Radjenović, D, Heričko, M, Torkar, R, & Živkovič, A. (2013). Software fault prediction metrics: A systematic literature review. *Information and Software Technology, 55*(8), 1397–1418.
4. He, Peng, Li, Bing, Liu, Xiao, Chen, Jun, & Ma, Yutao. (2015). An empirical study on software defect prediction with a simplified metric set. *Information and Software Technology, 59*, 170–190.
5. Selvarani, R., & Mangayarkarasi, P. (2015). A Dynamic Optimization Technique for Redesigning OO Software for Reusability. *ACM SIGSOFT Software Engineering Notes 40* (2), 1–6.
6. IEEE Standard 1633[TM]. (2008): *IEEE recommended practice on software reliability.* IEEE Reliability Society, sponsored by the Standards Committee.
7. Mohanta, S., Vinod, G., Mall, R. (2011). A technique for early prediction of software reliability based on design metrics. *International Journal of System Assurance Engineering and Management, 2*(4), 261–281.
8. Wu, H. L., Zhong, Y., & Chen, Y. (2011). A software reliability prediction model based on benchmark measurement. In *IEEE International Conference on Information management and Engineering* (pp. 131–134).
9. Stineburg, J., Zage, D., & Zage, W. (2005). Measuring the effect of design decisions on software reliability. In *International Society of Software Reliability Engineers* (ISSRE).
10. Eusgeld, I., Fraikin, F., Rohr, M., Salfner, F., & Wappler, U. (2008). Software reliability (pp. 104–125). Heidelberg: Springer.
11. Roshandel, R., Banerjee, S., Cheung, L., Medvidivic, N., & Golubchik, L. (2006). Estimating software component reliability by leveraging architectural models. In *ICSE'06, May 20–28.* Shanghai, China: ACM.
12. Xu, P., & Xu, S. (2010). A reliability model for object-oriented software. In *19th IEEE Asian Test Symposium.*
13. Schneidewind, N. F. (2004). A recommended practice for software reliability. *The Journal of Defense Software Engineering.* Cross Talk.
14. Kumar, K. S., & Misra, R. B. (2008). An enhanced model for early software reliability prediction using software engineering metrics. IEEE Computer Society.
15. Bharathi, R., & Selvarani, R. (2015). A framework for the estimation of OO software reliability using design complexity metrics. In *2015 International Conference on Trends in Automation, Communications and Computing Technology (ITACT-15)* (pp. 1–7). Bangalore.

16. ISO/TEC 9126-1(new): Software engineering-product quality: Quality model.
17. Selvarani, R., Gopalakrishnan Nair, T. R., & Kamakshi Prasad, V. (2009). Estimation of defect proneness using design complexity measurements in object-oriented software. In *Proceedings of International Conference on Computer Design and Applications(ICCDA) & (ICSPS)* (pp. 766–770). Singapore: IEEE Computer Society Press, CPS.
18. Gopalakrishnan Nair, T. R., & Selvarani, R. (2010). Estimation of software reusability: An engineering approach. *ACM SIGSOFT 35*(1), 1–6.
19. Gopalakrishnan Nair, T. R., Aravindh, S., & Selvarani, R. (2010). Design property metrics to Maintainability estimation—A virtual method using functional relationship mapping. *ACM SIGSOFT Software Engineering Notes 35*(6), 1–6.
20. Basili, V. R., Briand, L., & Melo, W. (1996). A Validation of object oriented design metrics as quality indicators. *IEEE Transactions on Software Engineering, 22*(10), 751–761.
21. Chidambar, S., & Kemerer, C. (1994). A Metrics Suite for Object oriented Design. *IEEE Transactions on Software Engineering, 20*(6), 476–493.
22. Gopalakrishnan Nair, T. R., & Selvarani, R. (2011). Defect proneness estimation and feedback approach for software design quality improvement. *Elsevier journal for Information and Software Technology 54*(2012), 274–285.
23. http://satc.gsfc.nasa.gov/support/STC_APR98/apply_oo.html.
24. Sjoberg, D. I. K., Dyba, T., & Jorgenson, M. (2007). The Future of Empirical Methods in Software Engineering Research. In *IEEE Computer society, Future of Software Engineering (FOSE'07)* 0-7695-2829-5/.
25. Kitchenham, B. A., Pfleeger, S. L., Pickard, L. M., Jones, P. W., Hoaglin, D. C., Emam, K. El., et al. (2002). Preliminary guidelines for empirical research in software engineering. In *IEEE Transactions on Software Engineering 28*(8), 721–734.

Using System Dynamics for Agile Cloud Systems Simulation Modelling

Olumide Akerele

Abstract Cloud Systems Simulation Modelling (CSSM) combines three different topic areas in software engineering, apparent in its constituting keywords: cloud system, simulation and modelling. Literally, it involves the simulation of various units of a cloud system—functioning as a holistic body. CSSM addresses various drawbacks of physical modelling of cloud systems, such as time of setup, cost of setup and expertise required. Simulation of cloud systems to explore potential cloud system options for 'smarter' managerial and technical decision-making help to significantly eradicate waste of resources that would otherwise be required for physically exploring cloud system behaviours. This chapter provides an in-depth overview of System Dynamics, the most widely adopted implementation of CSSM. This chapter provides an in-depth background to CSSM and its applicability in cloud software engineering—providing a case for the apt suitability of System Dynamics in investigating cloud software projects. It discusses the components of System Dynamic models in CSSM, data sources for effectively calibrating System Dynamic models, role of empirical studies in System Dynamics for CSSM, and the various methods of assessing the credibility of System Dynamic models in CSSM.

Keywords Cloud computing · System dynamics · Simulation models

1 Introduction

Cloud computing provides a model for delivering information technology services, using resources that are accessed from the Internet via web-based tools and applications, rather than a traditional connection to an onsite server. An electronic device is granted access to information, provided it has access to the web—this is facilitated by cloud computing. The complete embodiment of the various inherent functional units to facilitate this, their interactivity and their structure is referred to

O. Akerele (✉)
Parexel International, Nottingham, UK
e-mail: Olumide.akerele@parexel.com

© Springer International Publishing AG 2017
A. Hosseinian-Far et al. (eds.), *Strategic Engineering for Cloud Computing and Big Data Analytics*, DOI 10.1007/978-3-319-52491-7_6

as a Cloud System. The simulation of such systems—involving its design, modelling and calibration—is referred to as Cloud Systems Simulation Modelling (CSSM).

CSSM combines three different topic areas in software engineering, apparent in its constituting keywords: cloud system, simulation and modelling. Each of the keywords is defined individually and put together to originate the definition of CSSM in this paper.

A model is the theory of behaviour representing the workability of all or parts of an actual system of interest, incorporating its salient features [1]. Evolving the definition of a model, a simulation model is the numerical evaluation of the prototype of a system of interest, with the aim of predicting the performance of the system in the real world. Simulation models impersonate the operation of a real system. Typically, simulation models are dynamic in nature and are used to demonstrate a proof of concept [2]. Developing a basic simulation model involves the identification of three main elements: components of the modelled system, dynamic interaction between the parts of the modelled system and the nature of the system inputs [3].

CSSM combines the unique capabilities and characteristics of cloud system, simulation and modelling to model a cloud system of interest and study the behavioural evolution of such modelled process systems. As such, CSSM provides a great platform for conducting experiments to test various hypotheses in cloud software engineering domain. A diagram of CSSM as a method of experimentation on a cloud system model is shown in Fig. 1.

The second cadre in Fig. 1 shows that experiments in cloud systems can be conducted in two ways: formal experimentation and cloud system model experimentation. Formal experiments involve conducting experiments in a controlled environment to test the validity of a hypothesis. Software engineering researchers

Fig. 1 Methods of Studying Cloud Systems

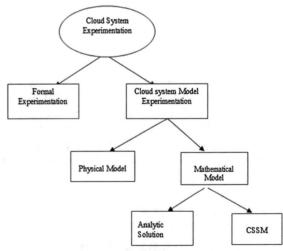

claim it is more desirable to investigate hypotheses in software projects by conducting formal experiments, as this reduces many validity concerns [4]. However, the implementation of formal experiments is limited by cost, time and other potential risks associated with empirical investigation of scientific hypotheses, particularly when dealing with; real complex and large projects [5, 6].

Weiss [7] confirms "... in software engineering, it is remarkably easy to propose a hypothesis and remarkably difficult to test them". For this reason, it is useful to seek other methods for testing software engineering hypothesis.

Experimentation on the system model plays a critical role in situations where formal experimentation on the actual system of interest is not achievable. The model of the developed cloud system may be a physical model, or more popularly, a mathematical model [8, 9]. A physical model involves building the physical prototype of the system to be investigated. Mathematical models on the other hand, represent software systems using logical and quantitative relationships which can then be applied to investigate various software system hypotheses.

Mathematical models applied in analytic solutions are used for static model representation, where simple relationships and quantities are adopted. However, this is not suitable for complex and dynamic cloud systems. CSSM addresses this concern by numerically evaluating cloud system models to determine how information inputs affect their output metrics of interest to answer various research questions in the cloud software engineering domain.

Simulation modelling was first applied in software engineering research by Abdel–Hamid [10] for his PhD, in which he adopted System Dynamics [11] in the investigation of various waterfall software projects related problems. This study revealed simulation models as a ground breaking methodology in software engineering and hundreds of publications have spiralled up since then, reporting research results on various problem domains of software engineering. As at 2012, Zhang et al. [12] collated various publications on simulation modelling spanning over 21 problem domains in software engineering.

As mentioned, CSSM is an apt substitute for formal experiments, particularly when the cloud system cost, time of design and setup, and system controllability are of essence. Armbrust emphasized on the cost advantage of CSSM, stating

> To get statistically significant results, a high number of experimental subjects is preferred. In addition to the costs for these subjects, the costs for experimental setup, conduction and subsequent data analysis must be considered. The enormous costs of experimentation have been noticed by other sciences years ago. Consequently, simulations were developed to save some of these costs [12].

By acting as a sandbox for carrying out various experiments without any ramifications, CSSM provides a more practical approach for investigating cloud system hypotheses when risks are high, and logistics of experimental setup are a challenge [13–15]. This makes CSSM suitable for investigating high risk process changes in cloud systems, and also for investigating the efficiency of new technologies.

2 Techniques of CSSM

The choice of CSSM technique adopted is dependent on the objective of the modeller, as well as the nature of the system to be investigated [16]. Of all the aforementioned CSSM techniques, continuous simulation, discrete-event simulation, and hybrid simulation are the three adopted techniques in software engineering. The majority of the other listed techniques are used solely for laboratory research and educational training [12].

System dynamic models a complex system using feedback loops and causal diagrams to monitor the system response continuously based on a unique set of equations. Discrete-event simulation technique on the other hand models the workability of a complex system as a discrete sequence of well-defined events. Hybrid simulation combines the characteristics of both continuous simulation and discrete-event simulation.

This paper focuses on the application of System Dynamics in CSSM. The next section provides in-depth details of System Dynamics and its application in CSSM.

3 Rationalization for Simulation Modelling in Investigating Cloud Software Project Problems

The effectiveness of various cloud software project factors lack uniformity across various software project environments due to the complex nature of cloud software projects [17]. The development and delivery of management systems consists of interactions and interdependencies that make it difficult to neither fully neither understand nor predict the behaviour of cloud software systems and practices. Abdel-Hamid [18] emphasizes on this major problem in predicting the output of software projects, particularly using empirical human analysis

> Software development processes and organizations are complex and it is therefore often hard to predict how changes made to some part of the process or organization will affect other parts and the outcome of the process as a whole …. In addition, like many processes, software processes can contain multiple feedback loops…. The complexity that results from these effects and their interactions make it almost impossible for human (mental) analysis to predict the consequences.

Further to Agarwal's statement, Zhang [12] also professes that "… software development is a highly complex, human-centered endeavour, which involves many uncertain factors in the course of the development process."

These statements suggest that cloud software projects are multi-faceted with various dynamic, complex and interrelated factors interacting in a continuous manner and creating revolving loops within software projects. As such, an overlooked variable in a cloud software project system may have a knock-on effect on other factors, leading to variations in the performance of factors of interest and consequently on the project performance. Such variations could be the thin line

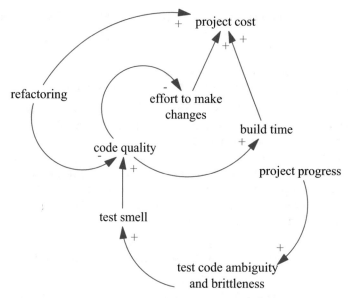

Legend: + denotes increasing effect; - denotes reducing effect

Fig. 2 Dynamic effects of code refactoring in cloud software development project

between the definition of the success or failure of software projects. Madachy [11] describes over 200 factors that affect the outcome of software projects.

A pragmatic example of a variable with dynamic and causal effects in cloud software project performance is the refactoring of a project test suite. This example is expressed diagrammatically in Fig. 2.

Figure 2 shows that as a cloud software project matures and automated acceptance test cases increase in direct proportion to the functionalities developed, the test suite complexity, brittleness, as well as coupling among the test cases increases. This leads to increasing ambiguity and brittleness in the test cases, leading to a gradual degradation in the design of the test suite known as test smell [19]. This is a depiction of poor code quality. The resulting test smell has been suggested in the literature to increase the build time required to run the acceptance test suite due to the ensuing increase in the redundancy of automated test cases [19].

Further to that, a cloud software project with poor test code quality is less maintainable and will require considerably more effort to introduce changes into it as a project matures when more functionalities are developed [19]. Both build time and the effort to make changes increase the project cost as they inhibit the progress of development. However, after refactoring of the test suite, a practice designed to improve the non-functional qualities of software, there is a significant drop in the test suite maintenance effort due to the improved design of the test scripts [19, 20]. Refactoring of course comes at a cost of extra project effort.

This example corroborates the understanding that every event has a cause which is itself a result of another cause, as exemplified by Sterman [21] using a popular micro-economics analogy

> Inventory is too high because sales unexpectedly fell. Sales fell because the competitors lowered their price. The competitors lowered their price because ... [21].

Investigating such complexity in cloud software project systems cannot be convincingly carried out with empirical techniques; hence, the need for cloud software project simulation modelling.

Furthermore, iterative and incremental approach of cloud software projects requires development teams to continuously develop software features incrementally and iteratively, which oftentimes builds up schedule pressure in the team when there is continuous accumulation of developed but undelivered features [22]. Hammond [22] summarized this occurrence

> While Agile speeds software design and development, it doesn't do much to speed up release and deployment—creating a flash point where frequent releases collide with slower release practices.

As a result, agile cloud software projects are particularly more vulnerable to schedule pressure, making the complexity of such systems even more pronounced and consequently exposing such systems to more project performance variations. Such queuing problems are more effectively dealt with using software methodologies such as Kanban and Lean which focus on Just-in Time (JIT) task scheduling.

Glaiel et al. [23] also claimed that schedule pressure is more prevalent in agile cloud based projects as a result of the short iterations of software delivery at the end of iterations, frequently interrupting the flow of continuity in development—compared to waterfall projects where such interruption only occurs once due to its big bang delivery approach. This view also strongly shared by Oorschot [24]. When these schedule pressure inducing forces described combine together, significantly inhibiting the productivity of the team, the complex effects of schedule pressure manifests in continuous delivery projects.

Empirical studies suggest teams cut down on the planned and standardized team development practices such as code reviews and agile practices that generally require extra effort to implement so as to boost their perceived productivity [25]. They take the shortest possible means to complete tasks so as to try and catch up with the lost work. In many cases, though the compromise of these standard practices may initially boost the productivity of the team, they end up increasing the defects discovered in the system and the required rework effort to fix these extra defects will come back to eventually reduce the productivity of the team.

The representation of the complexity of occurrences like schedule pressure and various other factors within software processes can only be effectively achieved using simulation modelling. Without simulation modelling, researchers will be forced to study software project systems as simple systems, ignoring the consideration of many underlying critical factors. As such, simulation modelling provides an evaluation platform for software process researchers to investigate modelled cloud software process systems—accounting for every interrelationship between the system variables and capturing all the behavioural intricacies in a way that is unachievable using any other empirical methods.

Finally, simulation modelling enables the full control of all the model variables in a cloud software delivery project. As such, 'noise' from unwanted variables during model experimentation could be obliterated by turning the variable off. Without this capability, it would be impracticable to effectively achieve the goals of cloud software project delivery research. Controlling perplexing or extraneous variables using empirical approaches such as formal experiments are very hard to achieve. This makes simulation modelling the most ideal choice to conduct similar studies in cloud software delivery projects.

3.1 Why System Dynamics?

As mentioned, three simulation techniques are in practical use: System dynamics (continuous simulation), discrete-event simulation and hybrid simulation. Evaluating the suitability and effectiveness of the three techniques, System Dynamics is deemed the most apt in agile cloud software projects for the reasons discussed in this section.

One major problem in agile cloud software projects is that the portrayal of the causes and effects of occurrences in software projects is hazy and the input and output of software project components are disproportional [26]. Also, the reaction to project manager's decisions is extremely time dependent, time-lagged and nonlinear in software projects [26]. There is a crucial need for support to represent and understand the underlying reasons why cloud software project systems behave the way they do. Perry et al. [27] summarized this concern elaborately in software engineering:

> … Software engineering research has failed to produce the deep models and analytical tools that are common in other sciences. The situation indicates a serious problem with research and practice in software engineering. We don't know the fundamental mechanisms that drive the costs and benefits of software tools and methods. Without this information, we can't tell whether we are basing our actions on faulty assumptions, evaluating new methods properly, or inadvertently focusing on low-payoff improvements. In fact, unless we understand the specific factors that cause tools and methods to be more or less cost-effective, the development and use of a particular technology will essentially be a random act.

System Dynamics addresses these concerns to justify its apt suitability for solving various cloud software project problems. Using its unique feedback loops, dynamic and nonlinear behaviours can be effectively represented and revealed. The use of System Dynamics helps to explore the complex dynamic consequences of various agile-based policies not achievable by any other simulation techniques [28]. This is because it facilitates the representation of the system as an integrative feedback system. Through the causal and feedback loops, System Dynamics helps to observe how the structural changes in one section of the modelled system may impact the holistic behaviour of the system.

Additionally, System Dynamics designs the cloud software project variables in the model as simultaneous activities, a capability not provided by other simulation techniques. This capability is vital in agile cloud software projects, particularly in scenarios where the continuous changes in various project variables need to be concurrently observed as the simulated project evolved.

Finally, as cloud software projects do not primarily focus on the individual activities and flow of the work processes, but rather on the quality of the work product in each activity, it will be wrong to model each process activity as a discrete-event; hence, the inappropriateness of discrete-event modelling technique. Observations of the continuous changes in the system are highly relevant in determining critical variables in the cloud software project system, particularly in determining the estimated schedule pressure.

4   System Dynamics

To understand the applicability of System Dynamics in software engineering research, *systems thinking*—the essence of system dynamics—is first explained.

4.1   Systems Thinking

Systems thinking is the knowledge of how the components of a system interact together to impact the behaviour of the holistic system is an area of particular interest [29]. In this context, a *system* refers to the combination of various parts organized for a specific purpose.

Systems thinking approach focuses on the practical knowledge of the (complex) interrelationships between the entire parts of a system, as against observation to understand the ways the parts function individually [30]. Daellenbach [31] succinctly describes systems thinking as a way of thinking in which "... *all things and events, and the experience of them, are parts of larger wholes ... this gave rise to a new way of thinking—systems thinking*". Systems thinking is further robustly defined as "*a conceptual framework with a body of knowledge and tools to identify wide-spread interactions, feedback, and recurring structures*" [11].

Drawn from various fields such as philosophy, sociology and organizational theory, systems thinking approach enables full system analysis and uncovers the majority of surprising and counter-intuitive outcomes of system behaviours compared to the simple and 'superficial' revelations of a basic cause-effect analysis. Such in-depth uncovering of the complex interrelationship in a system facilitates responsible decision making.

Systems thinking is based on four major concepts [11]:

- Thinking in circles and understanding that causes and effects are bidirectional.
- Envisaging system behaviour as a cause of elements within.
- Dynamic thinking rather than tactical thinking of interrelationships in the project system.
- A correlational versus operational orientation in investigating how effects actually do come to place.

4.2 System Dynamics Paradigm

System Dynamics is the computerized simulation execution of systems thinking in which all the state variables change continuously with time, as all the cause–effect relationships are modelled using continuous quantities. Madachy [11] further describes it as "*a methodology to implement systems thinking and leverage learning effort*".

The System Dynamics paradigm was developed in 1961 by systems engineer Jay Forrester, who applied it in management science [1, 32]. System dynamics enables the development of computer models representing complex systems and exploring various scenarios of the developed models to observe the overall system behaviour over time. Customarily, simulating such complex systems yields counter-intuitive behaviours that will normally not be revealed using static cause–effect analysis models that do not incorporate systems thinking. Sterman [33] describes the versatile applicability of system dynamics to any kind of active systems notable with mutual interaction, information feedback and interdependence.

Variables in System Dynamic models use mathematical equations to symbolize the various relationships among the model factors. This in fact facilitates the simulation capability of system dynamic models. A System Dynamic model has a nonlinear mathematical structure of first-order differential equations expressed as

$$y'(t) = f(y, m)$$

where y' represents vector of levels, f is a nonlinear function and m is a set of parameters.

Vensim is a free popular System Dynamics tool and has been adopted in previous studies in solving Software Engineering issues, including cloud software project issues [34, 15, 35]. It has the full capability to design System Dynamic models, and effectively run simulations over a specified period of time. In addition, it has the discrete-event modelling capability of introducing stochastic random behaviour into the model components.

4.2.1 Components of a System Dynamic Model

There are four major components of a system dynamic model: *level, rate, auxiliary, source* and *sink*. It is necessary to describe each of these components of to enable the understanding and interpretation of the developed System Dynamic models.

Level

Level, also known as *stock*, is the state variable that accumulates continuously within the simulation lifecycle. The terminology "level" was adopted to depict the accumulation of a variable over time, just as in the case of accumulation of liquid in a container over time. In general, a level provides information that forms the basis of decision making in System Dynamic models. The quantitative value of the level is the accumulation or the integration of the *rate* or *inflow* over a period of time. It is represented symbolically as a tub as shown in Fig. 3. Examples of levels in software projects include defects detected and fixed, expended project effort, project size, programmer experience [35].

Rate

Rate, also known as *flow*, is an action that directly changes the value accumulated in the level, positively or negatively. An *inflow* into a level increases the value of the level while an *outflow* reduces the value of the level. Usually, rates are functions of *auxiliaries*, *constants* and *levels*. It is represented by a stylized valve as seen in Fig. 4. For example, the productivity of the team is the *rate* impacting the size (in value) of developed software. Rates are referred to as control variables, as they are directly responsible for the value of levels in the system dynamic model. As such, rates represent the activities in a dynamic system.

Auxiliaries

Auxiliaries are variables formulated by the algebraic combination and computation of one or more information units. Usually, these information units are other variables and constants in the system. Typical examples include of auxiliaries include: defect density (function of defects detected and project size) and project progress (function of work completed and work remaining).

Source and Sink

The *source* and *sink* in a System Dynamic model represents a limitless exogenous supply and the repository of the *rate* respectively. They are exogenous in the sense

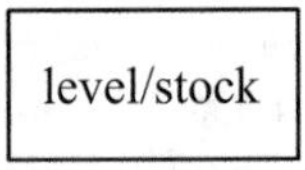

Fig. 3 Level

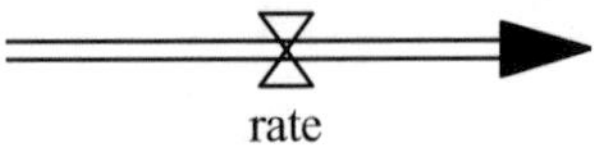

level/stock

Fig. 4 Rate

rate

Fig. 5 Source and sink

Fig. 6 An open-loop system

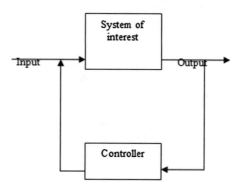

Fig. 7 A closed loop system

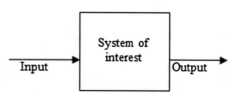

that they exist out of the model boundaries. Typical examples of software project source and sink are employee talent pool and employee attrition respectively. Sources and sinks are represented by a cloud symbol as shown in Fig. 5.

4.2.2 Components of a System Dynamic Model

An *open-loop system* is a system where the output is impacted by the input but not otherwise [30]. Consequently, the output does not have any effect on the input. Such systems neither receive nor give anything to the environment. The existence of actual open-loop systems in reality has been debated, with Daellenbach [31] arguing that the concept of an entirely open loop is theoretical and non-existent. He claims that systems have interactions with their environment, no matter how small the interaction is.

Conversely, a closed-loop system is identified as having the output influenced by its previous results, as it receives and/or supplies to its environment [31, 36]. As a result, the previous behaviour of a closed-loop system has an impact on the future behaviour of the same. Figures 6 and 7 represent the structure of an open and closed system.

The main significance of closed loops is that they cause dynamic counter-intuitive behaviour in systems. System Dynamics approach adopts closed loops, assuming that all systems consist of numerous closed-loop feedback and nonlinear effects.

4.2.3 Modelling and Simulation

The components of a system dynamics form the basic construct for developing a system dynamic model. Figure 8 represents a generic system dynamics structure. The basic structure consists of a stock and two rates portraying the two bidirectional flows for effectively accumulating and draining the stock.

The stock in Fig. 8 above retains the value of the inflow minus the outflow. Mathematically,

$$Stock(t_2) = \int_{t_1}^{t_2} [inflow(s) - outflow(s)]ds + Stock(t_1),$$

where t_2 and t_1 represent different times, and $t_2 = t_1 + 1$ time step.

The derivative of the change in the stock is represented by the following differential equation which all stocks are made of

$$\frac{\partial}{\partial t}(stock) = inflow(t_2) - outflow(t_2)$$

4.2.4 Causal Diagram and Feedback Loops

The primary step in developing a System Dynamic model is the identification of the exogenous and endogenous factors within the system and defining the interrelationships among the identified variables [30, 37]. Using arrows and the polarity signs, modellers describe the direction of the impact of one variable on the other and the nature of the impact respectively. A positive sign indicates a reinforcing effect while a negative sign indicates an abating effect. This graphical representation of such systems is known as *causal diagrams and feedback loops* [38].

Tvedt and Collofello [39] refer to the feedback loop as the "most powerful" feature of System Dynamics. Richardson (1991) also recognizes the central role of causal diagrams and feedback loops, describing them as the "heart" of system dynamics. A positive loop, also called a reinforcing loop, reinforces the initial reaction in the system. In other words, a positive loop amplifies the discrepancy

Fig. 8 Generic flow process in system dynamics

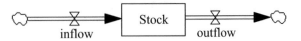

between a desired state and the future state of the system. Hence, positive loops are generally known to cause instability in a system and are undesirable. A negative loop or a balancing loop on the other hand counteracts the initial reaction in the system [30], i.e. it reduces the disparity between the desired state and future state of a system.

This high level graphical structure of causal and feedback loops, also called Qualitative System Dynamics, enhances a first glance understanding of complex systems [36]. Inter-connected feedback loops in a model often yield results that are counter-intuitive to what is reasonably expected when these loops are observed in isolation. At the model development stage, causal diagrams and feedback loops play two crucial roles: first, they help the modeller transform their underlying hypothesis into an elementary sketch. Secondly, they give a quick, reflective and effective overview a system to the stakeholders.

5 Data Sources

The effectiveness of a System Dynamic model to produce real-life results depends on the quality of the data sources used in the model development. The major sources of data for CSSM findings are from formal experiments, case studies, interviews and surveys [11, 34, 17]. In addition to this, data for System Dynamic model development can also be got from expert's subjective discernment and metrics from archived projects or estimates from personnel involved in the archived project [14].

6 Credibility of a Cloud Systems Simulation Model Using System Dynamics

The credibility of a simulation model relies on the success of a thorough verification and validation process [30, 36]. The purpose of model verification is to examine if the developed conceptual model accurately represents the structure of the real system while model validation examines if the accurately developed conceptual model correctly reflects the behaviour of the real system. They evaluate the "representativeness" and "usefulness" of the developed model [1]. Forrester [30] describes this process as a way of "establishing confidence in the soundness and usefulness of a model".

The verification and validation processes are critical in the development of meaningful cloud system models, as the model findings and conclusions are entirely dependent on the rightness of the qualitative and quantitative formulated assumptions underlying the structure of the model. If the fundamental assumptions used in the model development are wrong, all conclusions resulting from the simulation model will be misleading.

6.1   Cloud System Simulation Model Verification

A cloud system model structure is regarded as the driver of the system behaviour; thus, the critical need to ensure accuracy of the conceptual model structure (Harrell et al. [2]). Model verification establishes if the conceptual model structure accurately represents the system being studied. Generally, this process also involves model debugging to detect and correct any data errors in the model. This process is also popularly referred to as *structural validation* [30], *Perceived Representativeness of Models*—(PROM) [40] and *internal validity* [31].

6.2   Cloud System Simulation Model Validation

As mentioned, the validation phase tests the behavioural output of a simulation model is capable of behaving like the real modelled system [30]. It is also referred to as *behavioural validity* [2], *Analytical Policy of Quality Insights*—(AQ) [40] and *external validation* [31]. Various methods of testing the behavioural validity as suggested by Harrell et al. [2] are highlighted:

- Model comparison with an actual system.
- Testing for face validity by getting feedback from experienced and knowledgeable experts in the field on the reasonableness of the model behaviour.
- Evaluating model with historical data.
- Comparison with similar models.
- Testing extreme conditions.

6.3   Model and Actual Data Correlation for Validation Satisfaction

Researchers suggest that it is philosophically impossible to truly validate a model, adjudging that a simulation model can either be invalidated, i.e. shown to be wrong, or fail to be invalidated [31] Box and Draper [41] also stated "… *that all models are wrong; the practical question is how wrong do they have to be to not be useful*". This consensus was also summarized by Forrester regarding System Dynamic model validation, stating:

There is no way to prove validity of a theory that purports to represent behaviour in the real world. One can achieve only a degree of confidence in a model that is a compromise between adequacy and the time and cost of further improvement [42].

When the simulation output of a cloud system model is compared with that of a real system to test for validity, the level of mimicry needed to 'validate' a model is not standard. A 'perfect' correlation or exactness in the behavioural patterns of both the observed data and the simulation data is unachievable (Harrell et al. [2]). The main purpose of a simulation model is to examine "that the model's output behaviour has sufficient accuracy" for the model's intended purpose over the domain of model's intended applicability [43].

As such, a 'comforting' level of correlation between the simulation model and actual system output is the expectation for mode validation.

According to Daellenbach [31]: "what is or is not a close enough approximation is largely a question of judgement ... it depends on the precision needed in the system being developed". Thus, the 'comforting' level of correlation required is dependent on the degree of precision needed by the modeller.

7 Conclusions

This chapter has provided a comprehensive overview and applicability of CSSM as a research methodology paradigm in software engineering projects—focusing on the apt suitability of System Dynamics for investigating cloud software project related issues. System Dynamics, being the most suitable CSSM technique for many discussed reasons was zoomed upon, explaining the major components and notations in system dynamic models used in software engineering. This chapter also discusses the data sources for calibrating System Dynamic models. Finally, the various methods chosen for assessing the credibility of the CSSM were analyzed and discussed.

References

1. Forrester, J. W. (1961). *Industrial dynamics*. Cambridge, Massachusetts, USA: MIT Press.
2. Harrell, C., Ghosh, B., & Bowden, R. (2004). *Simulation using Promodel with CD-ROM* (2nd ed.). McGraw-Hill Science/Engineering/Math.
3. Zeigler, B. P. (2000). *Theory of modeling and simulation* (2nd ed.). San Diego: Academic Press.
4. Shull, F., Singer, J., & Sjøberg, D. I. K. (2007). *Guide to advanced empirical software engineering* (2008th ed.). London: Springer.
5. Jedlitschka, A., Ciolkowski, M., & Pfahl, D. (2008). reporting experiments in software engineering. In F. Shull, J. Singer & D. I. K. Sjøberg (Eds.), *Guide to advanced empirical software engineering* (pp. 201–228). London: Springer. Retrieved September 8, 2014, from http://link.springer.com/chapter/10.1007/978-1-84800-044-5_8.

6. Wohlin, C., Runeson, P., Höst, M., Ohlsson, M.C., Regnell, B., & Wesslén, A. (2012) Operation. In *Experimentation in software engineering* (pp. 117–122). Berlin: Springer. Retrieved September 7, 2014, from http://link.springer.com/chapter/10.1007/978-3-642-29044-2_9.

7. Weiss, D. M. (1984). Evaluating software development by error analysis: The data from the architecture research facility. *The Journal of Systems and Software, 1,* 57–70.

8. Law, A. M. (2006). *Simulation modeling and analysis* (4th ed.). Boston: McGraw-Hill Higher Education.

9. Akerele, O., Ramachandran, M., & Dixon, M. (2013). Testing in the cloud: Strategies, risks and benefits. In Z. Mahmood & S. Saeed (Eds.), *Software engineering frameworks for the cloud computing paradigm* (pp. 165–185)., Computer communications and networks London: Springer.

10. Abdel-Hamid, T. (1984). The dynamics of software project staffing: An integrative system dynamics perspective, Ph.D. dissertation, Massachusetts Institute of Technology.

11. Madachy, R. J. (2008). *Software process dynamics* (1st ed.). Piscataway, NJ, Hoboken, NJ: Wiley-IEEE Press.

12. Zhang, H. (2012) Simulation modeling of evolving software processes. In *2012 International Conference on Software and System Process (ICSSP)* (pp. 228–230).

13. Akerele, O., Ramachandran, M., & Dixon, M. (2014a). Investigating the practical impact of agile practices on the quality of software projects in continuous delivery. International Journal of Software Engineering (IJSSE), *7*(2), 3–38.

14. Munch, J., & Armbrust, O. (2003). Using empirical knowledge from replicated experiments for software process simulation: A practical example. In *2003 International Symposium on Empirical Software Engineering, 2003. ISESE 2003. Proceedings* (pp. 18–27).

15. Akerele, O., Ramachandran, M., & Dixon, M. (2014). Evaluating the impact of critical factors in agile continuous delivery process: A system dynamics approach, (IJACSA). *International Journal of Advanced Computer Science and Applications, 5*(3), 2014.

16. Korn, G. A. (2007). *Advanced dynamic-system simulation: Model-replication techniques and Monte Carlo simulation* (1st ed.). Hoboken, NJ: Wiley-Interscience.

17. Kellner, M. I., Madachy, R. J., & Raffo, D. M. (1999). Software process simulation modeling: Why? What? How? *Journal of Systems and Software, 46*(2–3), 91–105.

18. Abdel-Hamid, T., & Madnick, S. (1991). *Software project dynamics: An integrated approach* (1st ed.). Englewood Cliffs, NJ: Prentice Hall.

19. Fowler, M., Beck, K., Brant, J., Opdyke, W., & Roberts, D. (1999). *Refactoring: Improving the design of existing code* (1st edn.). Reading, MA: Addison Wesley.

20. Humble, J., & Farley, D. (2010). *Continuous delivery: Reliable software releases through build, test, and deployment automation.* Addison Wesley.

21. Sterman, J. (2000). *Business dynamics: Systems thinking and modeling for a complex world with CD-ROM* (p. 192). Boston, Mass, London: McGraw-Hill Higher Education.

22. Hammond, A., & Jeffrey, S. (2011). Five ways to streamline release management. Forrester Research Inc. Retrieved June 12, 2014, from http://www.serena.com/docs/repository/solutions/Forrester-Five_Ways_to_Streamline_Release_Management-from_Serena_Software.pdf.

23. Glaiel, F., Moulton, A., Manick, S. (2013). Agile dynamics: A system dynamics investigation of agile software development methods. Working Papers, Composite Information Systems Laboratory (CISL) Sloan School of Management, Massachusetts Institute of Technology, Cambridge.

24. Oorschot, K. E. (2009). Dynamics of agile software development. In *Proceedings of the 27th International Conference of the System Dynamics, July 26–30, Massachusetts, USA.*

25. Cohn, M. (2009). *Succeeding with agile: Software development using scrum* (1st ed.). Upper Saddle River, NJ: Addison Wesley.

26. Hughes, B., & Cotterell, M. (2005). *Software project management* (4th ed.). London: McGraw-Hill Higher Education.

27. Perry, D. E., Porter, A. A., & Votta, L. G. (2000). Empirical studies of software engineering: A roadmap. In *Proceedings of the Conference on the Future of Software Engineering, ICSE'00* (pp. 345–355). New York, NY, USA: ACM. Retrieved August 15, 2014, from http:// doi.acm.org/10.1145/336512.336586.
28. Kong, X., Liu, L., & Chen, J. (2011). Modeling agile software maintenance process using analytical theory of project investment. *Procedia Engineering, 24,* 138–142.
29. Blanchard, B. S. (2008). *System engineering management* (4th ed.). Hoboken, NJ: Wiley.
30. Forrester, J. W. (2013). *Industrial dynamics.* Martino Fine Books.
31. Daellenbach, H. G. (1994). *Systems and decision making: A management science approach.* Chichester. Madachy, R. J. (2008). *Software process dynamics* (1st ed.). Hoboken, NJ ; Piscataway, NJ: Wiley-IEEE Press; Wiley & Sons.
32. Armbrust, O. (2003). Using empirical knowledge for software process simulation: A practical example. Retrieved June 12, 2014 from, http://ove-armbrust.de/downloads/Armbrust-da.pdf.
33. Sterman, J. (2000). *Business dynamics: Systems thinking and modeling for a complex world with CD-ROM* (p. 192). Boston, Mass; London: McGraw-Hill Higher Education.
34. Akerele, O., & Ramachandran, M. (2014b). Continuous delivery in the cloud: An economic evaluation using system dynamics. In M. Ramachandran (Ed.), *Advances in cloud computing research.* Hauppauge, New York: Nova Science Pub Inc.
35. Akerele, O., Ramachandran, M., & Dixon, M. (2013a). System dynamics modeling of agile continuous delivery process. In *Agile Conference (AGILE), 2013* (pp. 60–63).
36. Coyle, G. (2000). Qualitative and quantitative modelling in system dynamics: Some research questions. *System Dynamics Review, 16*(3), 225–244.
37. Münch, J. (2012). *Software process definition and management* (2012th ed.). Heidelberg: Springer.
38. Richardson, G. P. (1991). System dynamics: Simulation for policy analysis from a feedback perspective. In P. A. Fishwick & P. A. Luker (Eds.), *Qualitative simulation modeling and analysis.* Advances in simulation (pp. 144–169). New York: Springer. Retrieved August 18, 2014, from http://link.springer.com/chapter/10.1007/978-1-4613-9072-5_7.
39. Tvedt, J., & Collofello, J. (1995). Evaluating the effectiveness of process improvements on software development cycle time via system dynamics modeling, Ph.D. thesis, University of Arizona.
40. Lane, D. C. (1998). Can we have confidence in generic structures? *The Journal of the Operational Research Society, 49*(9), 936.
41. Box, G. E. P., & Draper. (1987). *Empirical model-building and response surfaces* (1st edn.). New York: Wiley.
42. Forrester, J. W. (1994). System dynamics, systems thinking, and soft OR. *System Dynamics Review, 10*(2–3), 245–256.
43. Sargent, R. G. (1998). Verification and validation of simulation models. In *Proceedings of the 30th Conference on Winter Simulation. WSC'98, Los Alamitos, CA, USA* (pp. 121–130). IEEE Computer Society Press. Retrieved September 11, 2014, from http://dl.acm.org/citation.cfm? id=293172.293216.

Software Process Simulation Modelling for Agile Cloud Software Development Projects: Techniques and Applications

Olumide Akerele

Abstract Software Process Simulation Modelling has gained recognition in the recent years in addressing a variety of cloud software project development, software risk management and cloud software project management issues. Using Software Process Simulation Modelling, the investigator draws up real-world problems to address in the software domain, and then a simulation approach is used to develop as-is/to-be models—where the models are calibrated using credible empirical data. The simulation outcome of such cloud system project models provide an economic way of predicting implications of various decisions, helping to make with effective and prudent decision-making through the process. This chapter provides an overview of Software Process Simulation Modelling and the present issues it addresses as well as the motivation for its being—particularly related to agile cloud software projects. This chapter also discusses its techniques of implementation, as well as its applications in solving real-world problems.

Keywords Software process · System dynamics · Simulation models

1 Introduction

Software Process Simulation Modelling (SPSM) synergizes three different areas of expertise in software engineering, presented in the constituting keywords: *software process*, *simulation* and *modelling*. Each of the keywords is initially described to gain the full understanding of SPSM.

Sommerville [1] describes software processes—both plan-driven and agile—as structured activities identified to complete the development of software products. Paulk [2] gives a more representative description and defines software processes as "a set of activities, methods, practices, and transformations that people use to develop and maintain software and the associated products". Software processes

O. Akerele (✉)
Parexel International, Nottingham, UK
e-mail: Olumide.Akerele@parexel.com

© Springer International Publishing AG 2017
A. Hosseinian-Far et al. (eds.), *Strategic Engineering for Cloud Computing and Big Data Analytics*, DOI 10.1007/978-3-319-52491-7_7

occur from the inception of a software project when project requirements are discussed up until maintenance activities of the software. Standard software processes in software engineering include requirement specification, design, testing, inspections, deployment, etc. [3].

A model is the 'theory of behaviour' representing the workability of all or parts of an actual system, incorporating the salient features of the system [4]. It is the physical representation of a real-life system, possessing the significant characteristics of the system of interest.

Evolving the definition of a model, a simulation model is the numerical evaluation of the prototype of a system of interest, with the aim of predicting the performance of the system in real world [5]. It involves the impersonation of the operation of a real system. Typically, simulation models are dynamic in nature and are used to demonstrate a proof of concept [5]. Developing a basic simulation model involves the identification of three main elements: components of the modelled system, dynamic interaction between the parts of the modelled system and the nature of the system inputs [6].

SPSM combines the domains of 'software processes', 'simulation' and 'modelling' to produce software process models and study the behavioural evolution of the modelled systems [7]. Over the past decade, SPSM has grown and made a significant contribution in research of software engineering projects [8, 9].

SPSM was first applied in software engineering research by Abdel–Hamid for his PhD thesis [10], in which he adopted system dynamics in the investigation of various project policies and study the dynamics of waterfall software projects. This study revealed SPSM as a ground-breaking methodology in software engineering and hundreds of publications have spiralled up since then, reporting results on critical aspects of software engineering [7]. As at 2012, these publications span over 21 problem domains in software engineering, including agile software engineering, software evolution, requirements engineering, software evolution, etc. [11]. A trend analysis of most publications of in software project research using process simulation showed that the common simulation outputs are based on project duration, software quality and project cost (effort) and project size [11].

SPSM presents the simulated software development process in its currently implemented architecture (as-is) or in a proposed planned architecture for the future (to-be) [7]. Unlike static models, software process simulation models are executable and allow the behavioural description of the model elements. The modeller has the capability to make alterations to the model components and the inter-relationships between the model components give the system modeller the privilege of fully exploring the investigated system.

As mentioned, SPSM is a perfect substitute for experimentation on physical software process systems, particularly when project cost, time and controllability are constrained [12]. Munch and Armbrust commented on the cost advantage of SPSM:

> To get statistically significant results, a high number of experimental subjects are preferred. In addition to the costs for these subjects, the costs for experimental setup, conduction and

subsequent data analysis must be considered. The enormous costs of experimentation have been noticed by other sciences years ago. Consequently, simulations were developed to save some of these costs [12].

By acting as a sandbox for carrying out various experiments without any ramifications, SPSM provides a more practical approach for investigating software process hypothesis when risks are high, and logistics of experimental setup are a challenge. This makes SPSM suitable for investigating high risk process changes in software projects and also for investigating efficiency of new technologies [7, 12].

Although SPSM is a totally unique methodology, it is not an independently functional approach. The successful implementation of SPSM relies heavily on empirical studies. The major empirical research techniques in software engineering and the role of empirical studies in SPSM are hereby discussed.

2 Empirical Studies in Software Engineering

Empirical studies in software engineering are useful in examining various ideas, usually with the objective of investigating an unknown [13]. They are used in the comparison of what researchers observe to what they believe in, by examining the validity of a hypothesis via observation, either directly or indirectly ([13]. While empirical studies hardly give repeatable results, they yield evidence of causal relationships in specific settings. The investigator collates relevant qualitative and/or quantitative experimental data results and performs essential analysis on the data to derive answers to the research questions. Based on this results analysis, the investigator accepts or rejects the formulated hypothesis. In essence, measurement plays a vital role in empirical software engineering.

2.1 Empirical Approaches in Software Engineering

Empirical studies are broadly classified into qualitative and quantitative research methods, depending on the nature of data and purpose of the study [14]. In software engineering, the three main approaches of empirical studies are *formal experiment*, *case study* and *survey* [15].

2.1.1 Formal Experiment

A formal experiment uses statistical methods to test the validity of a hypothesis in a controlled and manageable environment [12]. In a formal experiment, one or more conditions—referred to as the *independent* variables—are altered to investigate their impact on one or more *dependent* variables. The data group observed after the intentional variable alteration is referred to as the *experimental group* while the data group observed prior to the independent variable alteration is referred to as the *control group* [15].

The formulated hypothesis—a pre-condition for an experimental approach—determines the relevant variables to be altered and observed in the experiment. A statistically significant result from the experiment should be sufficient to reject or be unable to reject the formulated hypothesis.

Formal experiments facilitate the full controllability of the system variables; hence, experimental results are highly replicable. However, to get a statistically significant result, a high number of subjects is required for a formal experiment; this may be significantly expensive. Also, the experimental setup and data analysis costs are deterrents to adopting formal experiments in software engineering [12].

2.1.2 Case Studies

A case study is the systematic description and analysis of a single case or a number of cases [16]. Yin [17] describes case study as an *empirical inquiry that investigates a contemporary phenomenon within its real-life context, especially when the boundaries between phenomenon and context are not clearly evident* [17].

Usually, multiple case studies are preferable over single case studies in software engineering so as to reduce the risk of misinterpretation and also to improve the generalisation and robustness of the research conclusions [18]. Although case studies are usually qualitative in nature, they may contain quantitative elements. Case studies are functional in addressing causal complexities, fostering new hypothesis and generally achieve high construct validity.

Case studies are useful in situations with multiple variables of interest, several sources of evidence and rich contexts with relatively high effectiveness in answering the 'how' and 'why' questions of research questions. Case studies have the benefit of not incurring the scaling problems associated with the formal experiments [16]. Also, the cost of carrying out case studies is relatively low due to the completion of the examined case.

A major downside of case studies is the lowness in the confidence and generalisation in the cause–effect relationships between investigated factors and results [18]. Case studies also involve a number of systematic techniques (i.e. data collection techniques, data analysis and data reporting) which may be challenging to efficiently put into practice [16]. Furthermore, case studies do not yield statistically significant results due to the small sample size of the examined project(s).

2.1.3 Survey

Survey is a retrospective research method in which data is collected from a broad population of individuals [12]. Generally, data is got from a target population in a structured format using questionnaires, interviews, telephoning, etc. Surveys are ideal for large number of variables with sample sizes—and also rely heavily on the extensive use of statistics for results analysis [19].

Surveys in software engineering are used to identify the target audience's reaction to tools, practices and new technologies in software engineering [19]. They are also effective in examining particular trends and investigating relationships between real-life processes. The key strength of surveys is the generalisation of the results to several projects and their relative ease of being understood and analysed.

A major disadvantage of surveys is that the researcher does not have control over the variables, due to their retrospective nature of its approach; consequently, there is low degree of certainty in the actual cause of an event. Also, bias of respondents could lead to inaccurate results and conclusion. Additionally, the development of rich, well-structured and detailed questionnaires could be extremely time-consuming.

Irrespective of the empirical study approach used, empirical research follows a sequence of five activities [13]:

- Formulating a hypothesis.
- Observing the situation of interest.
- Documenting data from observations.
- Data analysis.
- Conclusion got from analysed data.

Despite the popularity of adopting empirical studies in software engineering, it is reported to have had limited success compared to other fields [13]. Perry et al. commented:

> ... Empirical studies would still fail to have the impact they have had in other fields. This is because there is a gap between the studies we actually do and the goals we want those studies to achieve [13].

A major reason for this limitation is their inability to represent the full mechanism of the complexity of software projects [20]. It remains difficult to understand and study the underlying causes and effects of various occurrences in projects. This is the major contribution of SPSM in software engineering: to provide a pragmatic effective approach of carrying out research of software processes without the limitations of empirical techniques.

2.2 Role of Empirical Studies in Software Process Simulation Modelling

Empirical findings, as well as experience from real practices and processes are needed to effectively calibrate variables in SPSM; on the other hand, the model simulation results are also used to design and analyse real-life experiments [12]. In other words, the simulation input variables are obtained by observing a real project system while the simulation output variables are obtained by observing the real system from the simulation experiment [21].

This 'symbiotic' relationship suggests that although the use of SPSM results in empirical studies is optional, the incorporation of empirical results in SPSM is essential and mandatory for to achieve realistic results. Armbrust avowed the criticality of empirical studies in software process simulation modelling:

> One of the key problems of simulation is to create accurate models reflecting real world behaviour for specific contexts. Correctly mapping the real world to the model, executing the simulation, and mapping the results back to the real world is crucial to the success of the modelling and simulation effort. Integrating empirical knowledge into simulation is an important approach to achieve this goal, because it allows for the creation of a better image of the real world and a closer connection of real and virtual world... [22].

The trueness and credibility of a developed simulation model output is entirely dependent on the quality of the empirical data used for its structural development and parameterization [23]. If the empirical data input is floored, the simulation results will not replicate that of a real system [24]. On the other hand, high quality SPSM, the most popular source of reported empirical data used in model development is from formal experiment [20].

3 Agile Delivery of Cloud Projects

3.1 Rationale for Agile Development

Agile cloud software delivery practices advocate a unique set of practices for teams to achieve fluency in the frequent delivery of good quality software. Literally, software is part of the human day-to-day living, from embedded systems in wrist watches to critical health systems, from financial systems to social media applications, from word editing applications to road traffic systems. In 2014, the global market software revenue was estimated to be $321 billion, with over 17 million developers in the world [25]. Despite this success and pervasiveness of software systems, the development process has not been without severe challenges.

About 50 years ago, the term 'software crisis' was coined to portray the software industry's incapacity to provide customers with high quality software, developed within budget and on schedule [24]. Many software projects were called 'death march' projects as customers watched powerlessly how their software projects sank into the sea of failure [26]. Gibbs [27] claimed the average software development project overshot its schedule by half, with larger projects generally performing worse. He further suggested that three quarters of all large systems were 'operating failures' that did not function as intended. Sommerville [1] also faulted the performance of software projects, asserting that the majority of software failures were as a result of software failing to meet the user expectations. All the projects referenced by these authors had something in common: they were developed using the waterfall methodology [25].

The waterfall methodology is a rigid and sequential development process with strong emphasis on upfront design, requirements freeze and intensive planning. In waterfall projects, project advancement is measured in accordance to progression in the phases of the development process, as against the actual value of work completed. Accordingly, a change was needed to introduce a more adaptable development methodology, designed to improve the performance of software projects and cut down the project failure rate.

Between the 11th and 13th of February 2001, a group of 17 software consultants who referred to themselves as 'organisational anarchists' came together to address these fundamental software project issues and ended up proposing the agile software development methodology, a radical change to the popular waterfall methodology.

Agile software development adopts an incremental, iterative, adaptive, emergent and self-organising approach to developing software, with huge emphasis on satisfying customers by continuously welcoming requirements and delivering good quality software. At the end of this meeting, these software consultants signed the agile manifesto, enlisting the four core values of agile software development [28]:

- Individuals and interactions over processes and tools.
- Working software over comprehensive documentation.
- Customer collaboration over contract negotiation.
- Responding to change over following a plan.

Further to this, 12 principles to guide the implementation of agile software development were also agreed by these consultants. Over the years, various agile methodologies such as eXtreme Programming (XP), Scrum, Dynamic Systems Development Method (DSDM), Lean and Kanban have been originated with the ultimate goal of creating better software, all based on the fundamental values described in the agile manifesto. Results from a survey among 4000 software practitioners conducted to estimate the prevalence of agile software development adoption in the industry showed that over 84% of the respondents practised agile software development at their organisations [29].

The main difference between agile software development and the heavily structured waterfall methodology is that agile teams rely on the tacit knowledge possessed by their responsive teams, compared to waterfall project teams who stringently follow the step-by-step activities as specified in their full up-front documented plans. Although agile software development projects are prone to mistakes that could be discoverable during upfront planning in waterfall projects, waterfall projects bear the risk of severe cost if sudden changes in customer requirements occur, making their comprehensive and totally reliant upfront plans very expensive to update or completely antiquated [26, 30].

3.2   Agile Delivery Practices

Agile cloud software delivery practices advocate a unique set of practices for teams to achieve fluency in the frequent delivery of good quality software. These major practices include test automation, configuration management, continuous integration, deployment automation and DevOps. Figure 1 shows a cadre of delivery processes and their associated practices.

Figure 1 shows three different levels of software delivery, in an increasing order of automation level. Each delivery level adopts the practices of the preceding level in addition to its own practices. The practices at the bottom Agile delivery layer are typical of agile cloud software development projects such as the agile practices like pair programming test driven development, refactoring, iterative development approach, etc. These practices act as prerequisites to the other two paradigms above the agile movement layer.

The middle layer, continuous delivery, builds on the fundamental agile delivery approaches, proposing a generic set of practices including test automation, configuration management, continuous integration, deployment automation and DevOps to act as the backbone for the successful adoption of continuous delivery.

The subtle difference between continuous deployment and continuous delivery is in manual authorization for automatic deployment: while continuous delivery requires manual authorization to automatically deploy the latest version of a software into the customer's production environment after satisfactorily passing all associated tests, every successful build of the latest version of developed software is

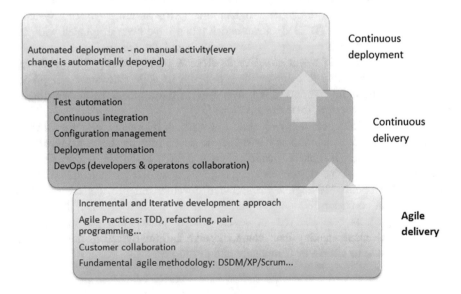

Fig. 1   Agile cloud software project delivery practices

automatically deployed without any human intervention in the case of continuous deployment.

4 Application of Software Process Simulation Modelling

SPSM is deployed for a number of uses in software development. Kellner et al. [7] classified the purposes of SPSM under six categories: strategic management, planning, control and operational management, process improvement and technology adoption, understanding, and training and learning. Zhang [11] further classified ten various purposes of SPSM into three areas: cognitive level, tactical and strategic levels. The purposes and applications of SPSM adoption have been classified under four major clusters in this chapter as shown in Fig. 2.

4.1 Strategic and Risk Management

SPSM is vital in key strategic decision-making to reduce or confer potential software project risks [31]. SPSM has been usefully deployed in supporting high level decision-making by project managers in taking the right directions during a software project. Software project management use SPSM to simulate and compare results of various scenarios, basing their decisions on the most favourable

Fig. 2 Purposes of SPSM

results [23]. SPSM is applied in strategic and risk management in answering questions like: "What degree of reuse should be present in the developed system?"; "What software components are better developed offshore?" [9]; "What impact will hiring graduate programmers instead of experienced programmers have on the schedule of the software project?" [8]; "Is 'team A' in location 'A' better suited for the development task than 'team B' in location 'B'?" [9]; "Is the project better outsourced or developed in-house?" [23].

4.2 Planning and Control

This is the commonest purpose of SPSM adoption [7]. SPSM is effectively used as a planning and control tool due to its effectiveness and precision in project scheduling and cost estimations [8, 11] It facilitates 'responsible decision making' as against managerial decision-making based on intuition [32]. As a planning and control tool, SPSM can answer critical questions like: "How long will the project take?" [8]; "What is the estimated project cost (in manpower)?" [31]; "How much personnel will be required for project completion?" [11, 23]; "How much impact will it have on the project if the developers are split across two simultaneous projects? [23]"; "How many developers will be needed to complete project 'A' in 7 months?" [23], etc.

4.3 Performance Analysis and Process Improvement

SPSM can be used to answer various questions relating to the investigation of the effectiveness of proposed improvement processes and practices; thus, acting as a handy tool for policymakers in introducing changes in their organisations [33]. It can be effectively used to 'pilot' changes in software processes before they are actually incorporated into the organisation and decisions could be made based on the impact of the various changes on the relevant project success metrics, usually the project cost, quality and schedule. The impact of changes to be investigated on project performance could be the effect of adopting new tools and technology, new 'quality improvement' practices, employing project manager with certain skills, etc.

Furthermore, SPSM is effective in supporting and moving organisations across all levels of the capability maturity levels, particularly helping organisations achieve capability maturity levels 4 and 5 via process definition and analysis, quantitative process management and continuous process improvement [23]. SPSM could be used to answer several questions like: "What is the impact on project schedule of testing software products in the 'cloud' compared to testing software products] on-site?" [33] "How cost effective is peer review?" [34, 35] "Does using tools for test automation improve software quality compared to manual testing?" [31], etc.

4.4 Experimental Learning and Training

Demonstrating software evolution for learning purposes to project stakeholders via simulations is a major application of SPSM, particularly when ideas cannot be demonstrated in real life due to cost or environmental constraints [7, 11]. Process flow of granulated activities within the software project system is also better explained, represented and understood using SPSM. The visual dynamic component representations enhance learning and also encourage exploration of the model.

Furthermore, SPSM can play the role of 'management flight simulator'—an analogy of the use of flight simulators by trainee pilots [24]. As such, SPSM can be used to demonstrate the impact of carrying out various decisions on the success of software projects to management project manager trainees. A popular lesson demonstrated is "adding people to a late project makes it later", first demonstrated by Brookes [36]. Demonstrating the failure of a project under such condition for example to project manager trainees is cheap, achievable and very insightful. New modellers are also encouraged to get quick hands-on experience, as SPSM requires little or no programming knowledge.

Although the model variable components utilise mathematical equations for simulations, they do not always require complex mathematical models. This makes SPSM relatively easy to use without any major learning or training [5]. SPSM can be used to demonstrate responses to questions like: "What will be difference in impact of adding five extra developers towards the end of development and adding them when the project is half done? [9]"; "How can the project be executed to be completed with two programmers in three months?" [23].

5 Techniques of Software Process Simulation Modelling

The choice of SPSM technique adopted is dependent on the aim and objective of the modeller, as well as the nature of the system to be investigated [7]. Primarily, SPSM is categorised based on three major criteria shown in Fig. 3.

SPSM can be either broadly classified as stochastic or deterministic in nature. Deterministic models adopt single values for the system variables and are executed only once. The model variable parameters are pre-determined due to high level of confidence in the sole value of the variable. On the other hand, stochastic simulation models execute the simulations over a defined number of times, using randomised parameters from a defined probability distribution for the simulated model variables [37]. Stochastic models are crucial in software projects for representing variables with high tendency of inconsistency across various environments, such as software project defect rate, productivity, attrition rate, etc.

Secondarily, SPSM may also be categorised as static or dynamic depending on the time-related nature of the system variables [38]. Static models are not time based and are inexecutable. They fundamentally reflect the snapshot of a system at

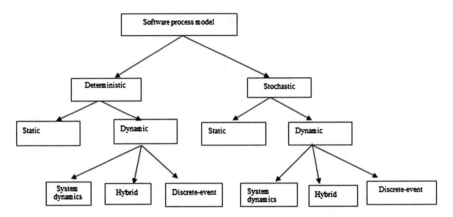

Fig. 3 SPSM techniques

a particular point in time. Conversely, dynamic models are time based, as the model state variables change over time. They are fully executable.

The third level of categorization of SPSM is based on the event handling approach of the model. Simulation techniques under this category include System Dynamics(Continuous simulation), Discrete-Event simulation, Hybrid simulation, Agent-Based Simulation (ABS), Role Playing Game Cognitive Map, Mode Driven Process Support Environments, Knowledge Based Systems and Multi-Agent Systems [39].

Of all these techniques, System Dynamics, Discrete Event simulation and Hybrid simulation are the three most popular and adopted techniques in software engineering; majority of the other listed techniques are used use solely for laboratory research and educational training [11]. A comprehensive systematic literature review also confirms that System Dynamics, Discrete Event and Hybrid simulation techniques are the most adopted SPSM paradigms [34, 35].

The next section briefly discusses the three common types of SPSM: System Dynamics, Discrete-Event and Hybrid simulation techniques.

5.1 System Dynamics

System Dynamics models a complex system using feedback loops and causal diagrams to monitor the system response continuously based on a unique set of equations. It facilitates the visualisation of the complex inter-relationship between variables in a software project system and runs simulations to study how complex project systems behave over time.

The System Dynamics paradigm was founded in 1961 by systems engineer Jay Forrester, where it was applied in management science [40]. However, System Dynamics was never applied in solving software project issues until

Abdel–Hamid's work in his PhD dissertation where system dynamics was used to study project dynamics and solve software project related issues [10]. A more detailed description and understanding of System Dynamics can be sought in [23].

More details on System Dynamics could be found in [5, 32].

5.2   Discrete-Event Simulation

Discrete-Event simulation technique on the other hand models the workability of a complex system as a discrete sequence of well-defined events. In Discrete-Event simulation models, the state of the system changes in response to the occurrence of specific events at discrete points in time [5]. Discrete-Event models have the capability to represent process structures and mechanisms to coordinate activities and convey work products while mainly focusing on the flow of discrete entities. Discrete-Event simulation models track individual events; unlike System Dynamics models where individual events are not tracked, focusing on system behavioural pattern. During time intervals in Discrete-Event simulation, the system state variables are constant.

Activities in Discrete-Event simulation models are represented in sequence, with the completion of each preceding activity triggering the commencement of the succeeding activity. Due to state changes only occurring following certain events, Discrete-Event simulation is useful in simulating detailed and fine-grained processes [32].

Models are calibrated with the actual unit of time of the real system. The simulation clock displays the time of the trigger of the next event during simulation run. Simulation runs are completed when there are no more events to occur or when the remaining duration falls short of the clock time.

A major benefit of Discrete-Event situation is its ability to develop stochastic models and effectively execute Monte Carlo simulations. In some cases, knowledge of programming is essential for the development of Discrete-Event simulation models.

5.2.1   Components and Terminologies

The major terms and components of a discrete event simulation model are the following:

1. *System*: Comprises of various substantial components called entities.
2. *Activities*: Points where there is change in action on the entities for a period of time. This may be designed using a probability distribution.
3. *System State*: The entirety of attributes or state variables describing the entities of the system.
4. *Queues*: Points where the model entities wait for an uncertain amount of time

5. *Entities*: Are tangible temporary or permanent elements that are acted on in activities and stay in queues.
6. *Simulation Clock*: An allocated variable in the model displaying the actual value of simulated time.
7. *Event*: An immediate incidence in time that usually alters the state of the system. The sequence of these events provides the dynamics of the simulation model.
8. *Event list:* Chronologically arranged list of events.
9. *Report generator*: Usually a subprogram in the model that calculates the approximations of the desired level of performance and generates a report when the simulation is completed.
10. *Event routine*: Usually a subprogram that updates system state of the model at the trigger of a specific event.
11. *Random number generator*: This is the pseudorandom number generator in the model to generate various kinds of random numbers from a determined probability distribution for the simulation model.
12. *Timing routine*: A subprogram that identifies the next event and fastens the simulation clock to the time its occurrence.
13. *Main programme*: A subprogram that summons the timing routing.

To execute the main programme, the simulation clock of the model is initialized to 0. This is followed by the initialization of the cumulative statistics to 0 and defining the initial system state, i.e. either queue empty or idle server. The occurrence time of the first arrival is then generated and placed in the event list, chronologically sorted. Following this, the 'next event' on the event list is then selected and the simulation clock advances to the time of this 'next event' and executes the event. If all the events on the event list have been executed, the simulation is concluded and a report is generated. If there are still events remaining, simulation returns to the 'next event stage'.

The most widely described Discrete-Event simulation scenario is in a bank system. In this case, Discrete-Event simulation can be used to resolve queuing problems such as determining the number of cashiers needed to serve a certain number of customers, based on the frequency of customers' arrival and average duration of the service required. These variables, duration of service and average frequency of customers are the randomised variables from a probabilistic distribution in the model. The recognisable entities in the system are customer 'queue' and 'cashiers'. Major events it the system are customer 'arrivals' and 'departures'. The system states are the 'number of customers' in the queue and the 'status' of the cashier, i.e. busy or idle.

5.2.2 Discrete-Event Simulation Model Development Procedure

Generally, the highlighted sequential steps below are necessary to build a running discrete-event simulation model.

- Identifying the system components, including all the activities of the system.
- Identifying the activities with queues.
- Mapping the activities to their essential entities.
- Deciding the queue discipline in the system. Examples of these are FIFO, Stochastic Fair Queuing (SFQ), Random Early Detection (RED), etc.
- Deciding the destination of entities on completion of activity.

5.3 Hybrid Simulation

Hybrid SPSM came into light in the 90 s due to the need to study and evaluate process dynamics at the micro and macro levels [7]. Although Hybrid simulation refers to the combination of two or more types of simulation techniques, it primarily refers to the combination of System Dynamics and Discrete-Event simulation paradigms in software process modelling. As such, Hybrid simulation technique in this chapter is a fusion of Discrete-Event simulation and System Dynamics.

The Hybrid SPSM paradigm designs cloud software project systems as discrete sequential units and executes them in a continuously changing environment [5]. Hybrid simulation models utilise the capabilities of both Discrete-Event and System Dynamics to model systems that are neither fully discrete nor continuous. The main idea behind the Hybrid process modelling technique is to breakdown discrete events into numerous iterations so as to capture the continuous changes of the model within one discrete activity.

The continuous simulation capability captures the dynamicity of the system environment while the Discrete-Event simulation feature captures the defined process steps. Table 1 below shows the summarised characteristics of both Discrete-Event simulation technique and System Dynamics.

Harrel et al. [5] describes three types of interactions that can exist between continuously changing and discretely changing variables in a simulation model:

- A continuous state variable may be influenced by the discrete change caused by a discrete event.
- The relationship administering a continuous state variable may be influenced by a discrete event.
- A discrete event may occur or be programmed as a result of a continuous state variable reaching its threshold value.

A more detailed description of Hybrid simulation technique could be found in [5, 32].

Table 1 Comparison of discrete-event models and system dynamics

Discrete event simulation	System dynamics
Runs faster; this is because the simulation time can be advanced to next event	Activities occurring in parallel can be represented and simulated simultaneously
Efficiently models process steps of a system activity	Facilitates the efficient update of continuously changing variables; not sole event times as in discrete event simulation
Ability to represent queues	Representation of complex inter-relationships among system components using feedback loops
The model entities have their distinct attributes	Captures the full dynamics of a system
Activity duration can be randomised; making room for uncertainty	Programming knowledge not essential
Enables the representation of the interdependence of activities in software projects	Graphical system presentation facilitates a one-glance high level understanding of causal effects of a system
Low level of information aggregation with sufficient process details	High level information aggregation

6   Sensitivity Analysis

Sensitivity Analysis, also known as *what-if analysis is* a technique used to observe the change in the behaviour of a system following the introduction of some uncertainty into the inputs of the simulated system. It is applied to test the robustness of a simulation model, check for model system errors and most importantly to facilitate system exploration to further understand the relationship between system variables.

In SPSM, uncertainty is introduced to the basic assumptions in the primary model so as to determine the impact of such uncertainties on the performance model variable outputs of interest [41]. As a result, a model is described as being 'sensitive' if an intentionally introduced change in a model results in a significant change in the performance variable output of interest.

Essentially, sensitivity analyses are a critical technique in SPSM research due to their pivotal role in model exploration. Fiacco [42] comments on the relevance of sensitivity analysis:

> …sensitivity and stability analysis should be an integral part of any solution methodology. The status of a solution cannot be understood without such information. This has been well recognised since the inception of scientific inquiry and has been explicitly addressed from the beginning of mathematics. [42].

In CSSM experimentation, sensitivity analysis is generally applied to further investigate the relationship between key variables so as to answer research questions of interest, usually after successful simulation model validation. There are four sequential steps in effectively conducting sensitivity analysis in SPSM:

- Identifying the variable inputs that will be perturbed and alter the values as required. Probability functions may be used.
- Identifying the output variables of interest for conducting regression analysis. Ideally, the output variable is related to the problem solved by the model.
- Running the simulation.
- Analysing the impact of the perturbation on the model output parameters.

There are four main types of sensitivity analysis in SPSM, depending on the objectives of the modeller: numerical sensitivity analysis, behavioural sensitivity analysis, structural sensitivity analysis and policy sensitivity analysis [23]. Numerical sensitivity analysis being concerned with the sensitivity of the model output variables of interest to an introduced parameter change in the model is the most popular sensitivity analysis technique used in SPSM research.

7 Literature Review

A historical note on the simulation of software projects is dated back to 1984 when Abdel–Hamid [10] applied system dynamics in solving software project related problems in his PhD thesis. Specifically, the author used system dynamics to solve software project staffing problems and demonstrate how system dynamics can be used to pilot various personnel staffing policies and determine the effect of these policies on software project performance. Simulation modelling was not applied to incremental software projects until 1998 when Collofello et al. [43] investigated the impact of unit testing on incremental software projects. Kuppuswami et al. [44] are credited with the first study on the application of SPSM in investigating the effectiveness of the 12 major XP practices in agile software projects. Strikingly, these three milestones in the application of simulation modelling to agile software projects are characterised by the use of system dynamics as the SPSM technique.

Other discovered studies based on the adoption of SPSM in agile development research using SPSM include studies on staff attrition [45], requirement volatility [46–48] task related parameters [49], iteration length [50], schedule pressure [51], Kanban and Scrum effectiveness [52] automation [53] and agile process [54] These studies are however not reported in this chapter as they do not specifically investigate any of the agile practices of interest.

8 Conclusions

SPSM has gained significant attention in the agile software development industry— mainly due to its economic edge of investigating the impact of major decisions on the success factors of cloud projects. This hinges on the prowess of conducting simulations on the evolution of project systems to envisage the impact of various actions on the key performance indicators of software projects. In the recent years in addressing a variety of cloud software project development, software risk management and cloud software project management issues. This chapter gives an extensive overview of SPSM and the problems it solves in the agile software development industry. This chapter also converses the various implementation techniques of SPSM in agile cloud software projects and its real-life application areas.

References

1. Sommerville, I. (2004). *Software engineering* (7th ed.). Boston: Addison Wesley.
2. Paulk, M. C. (1993) *The capability maturity model: Guidelines for improving the software process*. Addison-Wesley.
3. Paulk, M. C. (2001). Extreme programming from a CMM perspective. *IEEE Software, 18*(6), 19–26.
4. Forrester, J. W. (2013) *Industrial dynamics*. Martino Fine Books.
5. Harrell, C., Ghosh, B., & Bowden, R. (2004). *Simulation using promodel with CD-ROM* (2nd ed.). McGraw-Hill Science/Engineering/Math.
6. Zeigler, B. P. (2000). *Theory of modeling and simulation* (2nd ed.). San Diego: Academic Press.
7. Kellner, M. I., Madachy, R. J., & Raffo, D. M. (1999). Software process simulation modeling: Why? What? How? *Journal of Systems and Software, 46*(2–3), 91–105.
8. Akerele, O., Ramachandran, M., & Dixon, M. (2014a). Investigating the practical impact of agile practices on the quality of software projects in continuous delivery. *International Journal of Software Engineering (IJSSE), 7*(2), 3–38.
9. Akerele, O., Ramachandran, M., & Dixon, M. (2013a). System dynamics modeling of agile continuous delivery process. In *Agile Conference (AGILE)* (pp.60–63). UR.
10. Abdel-Hamid, T. (1984) The dynamics of software development project management: An integrative system dynamics perspective. [Ph.D. Dissertation] Sloan School of Management, MIT.
11. Akerele, O., Ramachandran, M., & Dixon, M. (2013). Testing in the cloud: Strategies, risks and benefits. In Z. Mahmood & S. Saeed (Eds.), *Software engineering frameworks for the cloud computing paradigm computer communications and networks* (pp. 165–185). London: Springer.
12. Munch, J., & Armbrust, O. (2003). Using empirical knowledge from replicated experiments for software process simulation: A practical example. In *2003 Proceedings of International Symposium on Empirical Software Engineering. ISESE 2003* (pp. 18–27).
13. Perry, D. E., Porter, A. A., & Votta, L. G. (2000). Empirical studies of software engineering: A roadmap. In *Proceedings of the Conference on the Future of Software Engineering. ICSE'00* (pp. 345–355). New York, NY, USA: ACM. Retrieved August 15, 2014, from http://doi.acm.org/10.1145/336512.336586.
14. Creswell, J. W. (2002). *Research design: Qualitative, quantitative, and mixed methods approaches (2nd ed.)*. Thousand Oaks, CA: SAGE.
15. Tichy, W. F., & Padberg, F. (2007). Empirical methods in software engineering research. In *29th International Conference on Software Engineering—Companion. ICSE 2007 Companion* (pp. 163–164).
16. Runeson, P. (2003). Using students as experiment subjects—An analysis on graduate and freshmen student data. *Proceedings of Seventh International Conference, Empirical Assessment and Evaluation in Software Engineering (EASE'03)*.

17. Yin, R. K. (2003) *Case study research: design and methods* (3rd ed.), Thousand Oaks, CA: Sage.
18. Flyvbjerg, B. (2006). Five misunderstandings about case study research. *Qualitative Inquiry, 12*(2), 219–245.
19. Punter, T., Ciolkowski, M., Freimut, B., & John, I. (2003). Conducting on-line surveys in software engineering. In *2003 International Symposium on Empirical Software Engineering. ISESE 2003 Proceedings* (pp. 80–88).
20. Wang, Q., Pfahl, D.,& Raffo, D. M. (2006). Software process change. In *International Software Process Workshop and International Workshop on Software Process Simulation and Modeling, SPW/ProSim.../Programming and Software Engineering* (2006 ed.). Berlin, New York: Springer.
21. Horton, G. (2001). Discrete-event simulation, introduction to simulation WS01/02-L04. Retrieved July 25, 2014, from http://isgwww.cs.uni-magdeburg.de/~graham/its_01/lectures/04-DESimulation-1.pdf.
22. Armbrust, O. (2003). Using empirical knowledge from replicated experiments for software process simulation: a practical example. In *2003 International Symposium on Empirical Software Engineering. ISESE 2003 Proceedings* (pp. 18–27).
23. Madachy, R. J. (2007). *Software process dynamics* (1st ed.). Hoboken, Piscataway, NJ: Wiley.
24. Abdel-Hamid, T., & Madnick, S. (1991). *Software project dynamics: An integrated approach* (1st ed.). Englewood Cliffs, NJ: Prentice Hall.
25. Statista. (2014). Worldwide IT spending on enterprise software from 2009–2015 (in billion U.S.dollars), The statistic portal. Retrieved September 9, 2014, from http://www.statista.com/statistics/203428/total-enterprise-software-revenue-forecast.
26. Cohn, M. (2005). *Agile estimating and planning* (1st ed.). Upper Saddle River, NJ: Prentice Hall.
27. Gibbs, W. W. (1994). Software's chronic crisis. *Scientific American* 72–81.
28. Agile Alliance. (2013). Guide to agile practices. Retrieved September 20, 2014, from http://guide.agilealliance.org/.
29. Version One. (2012), 7THAnnual State of Agile Development Survey. Retrieved July 27, 2014, from http://www.versionone.com/pdf/7th-Annual-State-of-Agile-Development-Survey.pdf.
30. Cohn, M. (2009). *Succeeding with agile: Software development using scrum* (1st ed.). Upper Saddle River, NJ: Addison Wesley.
31. Akerele, O., Ramachandran, M. & Dixon, M .(2014c). Evaluating the impact of critical factors in agile continuous delivery process: A system dynamics approach. *International Journal of Advanced Computer Science and Applications (IJACSA), 5*(3)
32. Daellenbach, H. G. (1994). *Systems and decision making: A management science approach.* Chichester: Wiley
33. Akerele, O., & Ramachandran, M. (2014b). Continuous delivery in the cloud: An economic evaluation using system dynamics. In M. Ramachandran (ed.), *Advances in cloud computing research.* Hauppauge, New York: Nova Science.
34. Lui, D., Wang, Q. & Xiao, J. (2009). The role of software process simulation modeling in software risk management: A systematic review. In *3rd International Symposium on Empirical Software Engineering and Measurement. ESEM 2009* (pp. 302–311).
35. Lui, K. M. & Chan, K. C. C. (2003). When does a pair outperform two individuals? In *Proceedings of the 4th International Conference on Extreme Programming and Agile Processes in Software Engineering XP'03* (pp. 225–233). Berlin, Heidelberg: Springer. Retrieved December 28, 2014, from http://dl.acm.org/citation.cfm?id=1763875.1763910.
36. Brooks, F. P. (1995). *The mythical man month and other essays on software engineering.* Addison Wesley.
37. Pinsky, M. A. & Karlin, S. (2010). *An introduction to stochastic modeling* (4th ed.). Amsterdam, Boston: Academic Press.

38. Velten, K. (2008). *Mathematical modeling and simulation: Introduction for scientists and engineers* (1st ed.). Weinheim, Germany: Wiley.
39. Scacchi, W. (2000). Understanding software process redesign using modeling, analysis and simulation. *Software Process: Improvement and Practice, 5*(2–3), 183–195.
40. Forrester, J. W. (1961). *Industrial dynamics*. Cambridge, Massachusetts, USA: MIT Press.
41. Menzies, T., Smith J., & Raffo, D. (2004). When is pair programming better? [Thesis] West Virginia University, Portland State University. Retrieved http://menzies.us/pdf/04pairprog.pdf.
42. Fiacco, A. V. (1983). *Introduction to sensitivity and stability analysis in nonlinear programming* (p. 3). New York.: Academic Press.
43. Collofello, J. S., Yang, Z., Tvedt, J. D., Merrill, D. & Rus, I. (1996). Modelling software testing processes. In *Conference Proceedings of the 1996 IEEE Fifteenth Annual International Phoenix Conference on Computers and Communications* (pp. 289–293).
44. Kuppuswami, S., Vivekanandan, K., Ramaswamy, P., & Rodrigues, P. (2003). The effects of individual xp practices on software development effort. *SIGSOFT Softwarare Engineering Notes, 28*(6), 6–6.
45. Collofello, J., Rus, I., Houston, D., Sycamore, D., & Smith-Daniels, D. (1998). A system dynamics software process simulator for staffing policies decision support. In *Proceedings of the Thirty-First Annual Hawaii International Conference on System Sciences,* Kohala Coast, USA (pp. 103–11).
46. Pfahl, D., & Lebsanft, K. (2000). Using simulation to analyse the impact of software requirements volatility on project performance. *Information and Software Technology, 42,* 1001–1008.
47. Al-Emran, A., Pfahl, D. & Ruhe, G. (2007). DynaReP: A discrete event simulation model for re-planning of software releases. Quoted in: Q. Wang, D. Pfahl, & D. M. Raffo (eds.). *Software process dynamics and agility*. Lecture notes in computer science (pp. 246–258). Berlin, Heidelberg: Springer. Retrieved August 29, 2014, from http://link.springer.com/chapter/10.1007/978-3-540-72426-1_21.
48. Al-Emran, A., & Pfahl, D. (2008). Performing operational release planning, replanning and risk analysis using a system dynamics simulation model. *Software Process: Improvement and Practice, 13*(3), 265–279.
49. Houston, D. X. & Buettner, D. J. (2013). Modeling user story completion of an agile software process. In *Proceedings of the 2013 International Conference on Software and System Process. ICSSP 2013* (pp. 88–97). New York, NY, USA: ACM. Retrieved August 20, 2014, from http://doi.acm.org/10.1145/2486046.2486063.
50. Kong, X., Liu, L., & Chen, J. (2011). Modeling agile software maintenance process using analytical theory of project investment. *Procedia Engineering, 24,* 138–142.
51. Oorschot, K. E. (2009). Dynamics of agile software development. In: *Proceedings of the 27th International Conference of the System Dynamics,* July 26–30. Massachusetts, USA.
52. Cocco, L., Mannaro, K., Concas, G. & Marchesi, M. (2011). Simulating Kanban and Scrum vs. Waterfall with System Dynamics. Quoted in: A. Sillitti, O. Hazzan, E. Bache, & X. Albaladejo (eds.), *Agile processes in software engineering and extreme programming*. Lecture notes in business information processing (pp. 117–131). Berlin, Heidelberg: Springer. Retrieved August 30, 2014, from http://link.springer.com/chapter/10.1007/978-3-642-20677-1_9.
53. Sahaf, Z., Garousi, V., Pfahl, D., Irving, R., & Amannejad, Y. (2014). When to automate software testing? Decision support based on system dynamics: An industrial case study. In *Proceedings of the 2014 International Conference on Software and System Process. ICSSP 2014* (pp. 149–158) New York, NY, USA, ACM, Retrieved January 9, 2015 from http://doi.acm.org/10.1145/2600821.2600832.
54. White, A. S. (2014). An Agile Project System Dynamics Simulation Model. *International Journal of Information Technologies and Systems Approach, 7*(1), 55–79.
55. Zhang, H. (2012) Simulation modeling of evolving software processes. In 2012 *International Conference on Software and System Process (ICSSP)* (pp. 228–230).

56. Münch, J. (2012). *Software process definition and management* (2012th ed.). Heidelberg: Springer.
57. Pritsker, A. A. B. (1995). *Introduction to simulation and SLAM II* (4th ed.). New York: Wiley.
58. Humble, J. & Farley, D. (2010) *Continuous delivery: reliable software releases through build, test, and deployment automation*, Addison Wesley.

Adoption of a Legacy Network Management Protocol for Virtualisation

Kiran Voderhobli

Abstract Virtualisation is one of the key concepts that allows for abstraction of hardware and software resources on the cloud. Virtualisation has been employed for all aspects of cloud computing including big data processing. There have been arguments based on recent research that indicate that computational efficiency could be more efficient via virtualisation compared to their physical counterparts. A data centre not only represents physical resources but also the collection of virtualised entities that in essence, form virtual networks. Methods to monitor these virtual entities for attributes such as network traffic, performance, sustainability, etc., usually tend to be deployed on ad hoc basis. Understanding the network related attributes of virtualised entities on the cloud will help take critical decisions on management activities such as timing, selection and migration of virtual machines (VMs). In corporate physical data networks, it could have been achieved using four of the network management functional areas, i.e., performance management, configuration management, accounting management and fault management. This chapter discusses, with examples, at how network management principles could be contextualised with virtualisation on the cloud. In particular, the discussion will be centred on the application of Simple Network Management Protocol (SNMP) for gathering behavioural statistics from each virtualised entity.

Keywords Network management · Virtualisation · Performance monitoring · SNMP · Cloud computing · Traffic characteristics · QoS

K. Voderhobli (✉)
School of Computing, Creative Technologies and Engineering,
Leeds Beckett University, Leeds, UK
e-mail: K.Voderhobli@leedsbeckett.ac.uk

© Springer International Publishing AG 2017
A. Hosseinian-Far et al. (eds.), *Strategic Engineering for Cloud Computing and Big Data Analytics*, DOI 10.1007/978-3-319-52491-7_8

141

1   Introduction

With the increase in cloud computing paradigm, as a basis for utility computing there is heavy reliance on resource sharing using virtualisation. Virtualisation is the underpinning concept that allows for abstraction of resources between various users (or subscribers) on an ad hoc basis. Virtualisation technologies have evolved rapidly in the recent years to represent more than just mere containers for operating systems. Today, the concept of virtualisation also includes virtualised network hardware, for example in the form of Infrastructure as a Service (IaaS). Modern businesses are moving away from hardware proliferation in favour of subscribing cloud-based virtual hardware. Hardware proliferation is expensive, difficult to manage and problematic to scale when demands change. One of the features of cloud computing is "on-demand-self-service" [1]. This allows businesses to increase or decrease access to virtualised entities based on ever-changing requirements and at a reduced cost. In most cases, cloud-based services are handled in the same way as in traditional systems in terms of usability.

Irrespective of which variant or adaptation of cloud computing model an organisation adopts, there is always a clear requirement to be able to account and manage the virtualised entities. As with all IT systems, cloud infrastructures need to be managed in such a way that the end users are shielded from the intricacies of management. The challenge facing modern network managers is ensuring that the virtualised entities are managed in the same way as real entities whilst recognising that the logical resources exist in a virtual plane. There is a need to manage both the physical resources and virtualised instances hosted by them. Furthermore, a network manager might need to collect network behavioural data from virtualised entities with the intent of feeding such data into decision-making—such as when to migrate a Virtual Machine (VM) and choosing the conditions that warrant migration.

Cloud computing involves deployment and use of large scale distributed resources. Given that modern distributed networks are error prone, the activities that encompass cloud management need to take a proactive approach rather than "fire-fighting". This is essential to ensure there are no issues with service availability to end users.

One of the problems facing modern network managers is with regard to employing typical "rule-book" de facto network management skills to cloud-based resources and applications. There is a dilemma on how to contextualise traditional LAN-based network management activities to large scale distributed and virtual entities. This chapter discusses how an established network management protocol could be employed in cloud-based contexts. The discussion revolves around the possibility of porting Simple Network Management Protocol (SNMP) to manage virtualised entities like virtual machines (VMs) and virtual routers (VR). This chapter provides an insight into current trends in using SNMP for virtualisation on the cloud. It highlights different perspectives, propositions, ideas and models put forward by the researching community and aims to give the reader a view on using SNMP network management techniques for cloud infrastructures.

2 Traditional Approach to Network Management Using SNMP

The issue with many of the standardised networking protocols founded in the era of TCP/IP is that they have not caught up with technological developments. Many protocols that were used as "stop-gap" interim solutions had been rolled out due to adoption by networking communities. This is also true with SNMP (and CMIP) which was standardised a few decades ago and designed for management of LANs. Corporate LANs were mostly contained locally and hence were easier to manage. In addition, legacy networks were created based on best-effort service, which is no longer the case for cloud-based networking that demand guarantee of delivery and low latency [2]. Management of entry/exit points, users and applications was easy. SNMP was never designed for the scale of computing, convergence and distributed nature we are dealing with today. In traditional organisational LANs, a network manager monitors a set of nodes and devices using a management station. The management station is able to poll devices for specific information on its network attributes. On receiving a manager's query, a process on an end device responds to the query. The management platform is able to process these results and present the network manager with useful results that helps to act upon changing network conditions or adverse events.

The SNMP protocol was first defined in RFC1213 to enable management of all networked devices [3]. SNMP follows a client/server approach with queries and responses sent to SNMP port 161. Any device that is configured to respond to a network manager's queries is a managed entity. Each managed entity maintains a Management Information Base (MIB), which records real-time values that represent events occurring at that managed entity. For example, an MIB records the number of TCP segments received, packets dropped, system up-time, local ports, routing tables, etc. From time to time, a network manager has to identify a set of devices on the network and the MIB objects to be queried to get a full picture on the status of devices and Quality of Service (QoS) on the network. MIBs record thousands of objects, some static and many dynamic, which represent the live behaviour of the device.

Over the years, network managers had to adapt to rapid and radical networking innovations, including virtualisation and cloud computing. With the advent of cloud computing there is no denying that there is the issue of managing networks that have no boundaries. Also prevalent was the problem of monitoring unlimited number of virtualised network entities in a very dynamic cloud environment. There were no recommendations on how to deal with these issues. The result was that the research community started proposing and implementing bespoke solutions to cloud network management. The networking community has been adapting to changes by creating platforms using older protocols to be used in modern contexts.

3 Virtualisation Network Attributes and IaaS

Virtualisation is the process of creating virtual instances by emulating real resources based on user requirements [4]. These resources could be Software as a Service (SaaS), Platform as a Service (PaaS) and Infrastructure as a Service (IaaS). IaaS is applicable to virtualisation of network resources where a user could subscribe to a set of network devices on demand, with access to a virtual network. Network Virtualisation allows for better utilisation of resources, reduces risks, reduces costs of ownership and increases fault tolerance [5]. Furthermore, in a large data centre, network virtualisation can help provide better management of interconnected servers [6]. Network Virtualisation is not a new concept as network engineers have been using virtual LANs for many years to provide network abstraction. However, the demand for network virtualisation has grown significantly most recently with the advent of cloud computing [7]. This has resulted in development of new networking techniques like software defined networking (SDN) and network function virtualisation (NFV).

The building block of network virtualisation lies in the substrate. The substrate is a collection of resources like network hardware, connections, networking software, etc. This collection is not necessarily a complete network but rather entities that could be instanced to create virtual networks. The layer above the substrate is the virtualisation layer that is responsible for creating logical resources from physical ones (in the substrate), for example, creation of a logical switch from a physical switch. The logical resources are used to create virtual resources (or virtual networks).

In IaaS, the actual physical resources are normally self-contained devices that have management features, meaning a network manager could gather statistical data from these devices. Any virtual instance can go through a cycle of various stages including being suspended, reallocated, migrated or any other form of consolidation process that might arise due to changing conditions. It is worth noting that there is a possibility that virtualised infrastructures could perform worse than their physical counterparts due to the same software being used to virtualise the server and the LAN hardware [8]. This is something cloud designers must consider as it could have implications on QoS.

4 Importance of Traffic Characteristics of Virtualised Network Devices

It is important to understand the rationale behind gathering network traffic characteristics of virtualised devices; and why this information is vital to a network manager. All devices, be it virtual or physical, exhibit certain "personalities" over the course of operation. Each of these entities has a network footprint which can be represented using a range of values including utilisation, idle times, data rate, number of connections, packet loss, etc. These values could feed into critical decision-making on performing maintenance on the virtualised devices (like migration, suspension, etc.).

4.1 QoS Requirements on the Cloud

Cloud Service Providers must ensure that sufficient QoS in accordance to Service Level Agreements (SLA) must be made available [9]. An IaaS infrastructure must provide efficient Quality of Service (QoS) irrespective of how much it scales in terms of number of users. For example, a white paper by Arista [10] highlighted low-latency switching and resilient networking as two of the characteristics of Virtualisation-optimised cloud networks. Low-latency switching is to ensure that bandwidth critical cloud applications are not hindered by poor latency. Resilient networking is also important for avoiding disruptions to workloads. These types of QoS attributes can only be guaranteed by active monitoring. In one of the recent discussions of challenges for cloud systems, it was highlighted that QoS-based resource allocation and provisioning must be considered. It was also mentioned that it is one of the concepts, which can be addressed at the level of management of the overall system [11]. A QoS aware cloud deployment should consolidate virtualised devices in such a way that performance for end users is optimum. This is where SNMP can prove to be useful as a traffic characterisation tool which could feed into a knowledge base.

4.2 Failure Forecasting

Cloud physical resources cannot be assumed reliable and tolerant to failures. A white paper by Cisco, Josyula et al. [12] stressed the importance of detecting abnormal operations in network infrastructures by using a range of methods including polling, SNMP traps and syslogs. Just as any network resource, cloud-based physical devices could fail unpredictably, thus having a knock-on effect on virtualised instances they represent. Hence, failure-aware resource allocation has also been mentioned as one of the consolidation schemes on the cloud.

Jailani and Patel [13] say that fault management can be broken down into four categories. These are detection, isolation, correction and administration. But it can be argued that applying these into traditional paradigms of failure forecasting using partly automated Network Operations Centre (NOC) would not be sufficient when applied to network function virtualisation (NVF) due to the need to orchestrate management of both virtual and physical resources. Gone are the days when a network manager could survey an entire network just by manual polling as virtualisation increases the scale of the managed domain. Therefore, it is paramount that the categories described by Jailani and Patel [13] are incorporated into autonomic management systems. Autonomic failure forecasting also achieves better resilience in cloud resources. Resilience is the ability of a system to recover from failures automatically based on analysis of various conditions and network traffic features. A high degree of resilience is required to ensure that service availability and QoS are not compromised during network disruption. This is also true with virtualised networks. In an interesting review on resilience in cloud computing, Colman-Meixner et al. [14] highlight the need for resiliency in cloud networks. They have described some techniques to achieve resilience in networks. Some of these include:

- Distribution of traffic and load to ensure that network traffic is allocated to different nodes to avoid loss due to failure of resources.
- Route replication to duplicate routes to mitigate the effect if link and node failures.
- Failure-aware provisioning to allocate resources based on predictions of possible faults.
- In the case of virtualised networks, the scheme to augment topologies where links and nodes are added to protect the system from failures.
- Traffic disruption avoidance for virtual networks to avoid known problem "hotspots".

For autonomic resiliency activities such as above, one can imagine the importance of maintaining an active knowledge base for failure probability detection based on traffic characterisation. A proactive network management polling scheme could alleviate problems caused by failure of devices by forecasting physical device behaviour based on traffic characterisation [15].

4.3   Move Towards Autonomic Management

In a widespread heterogeneous cloud environment, it is not feasible for a network manager to monitor each device on a 1:1 basis. Doing so would be onerous and too slow in gathering any meaningful live network information. Autonomic computing reduces human involvement by exhibiting self-management, self-healing and self-configuration properties [16]. To facilitate an autonomic environment, decisions are taken using a knowledge base created using run-time statistics. In the context of Cloud Network Management, values collected by SNMP polling could feed into decision-making for autonomic virtualisation management. It is also key to note that SNMP-based autonomic virtualisation management will fit perfectly with MAPE-K model. MAPE-K stands for Monitor, Analyse, Plan and Execute based on a knowledge base [17]. Kephart and Chess explain that knowledge is fundamental to autonomic behaviour. The knowledge base could be built from various sources including system configuration, log files and data gathered from sensors. The MAPE-K model is illustrated below in Fig. 1.

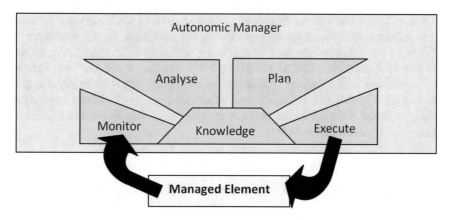

Fig. 1 MAPE-K model [17]

The elements of MAPE-K model can be contextualised into virtual network management as follows:

- The managed entity: This could be hardware or software. In the context of virtualised network management, the managed entities are network devices that can be virtualised.
- Sensors: to collect data that feed into the knowledge base. In the current context, this will be the SNMP agents, which are able to read MIB values. It is worth recalling from network management principles that each managed network device will have an agent.
- Effectors: to apply any changes to managed entities. In the current context, changes to network devices and virtualisation related operations could be initiated via APIs in the hypervisor. Alternatively, some operations to devices could be initiated by writing into MIBs.
- Autonomic Manager: to manage the virtual elements. If the human network manager is removed from the system, an autonomic manager could be implemented through the hypervisor (or even the substrate).
- Analysis and planning: Study the knowledge gathered continually to detect adverse conditions, changes, behaviour and thresholds. This can be implemented using software components deployed in the hypervisor.

Having looked at the rationale for active management of virtual devices, it would be appropriate to consider the feasibility of SNMP monitoring in a cloud environment. If a network manager were to use SNMP to manage virtual network devices, he would still have to follow the rules of SNMP communication—just as with traditional LANs. This would mean that a network manager should be able to perceive all the heterogeneous virtualised network devices as if they were physical. In simple terms, he should be able to query a virtual network device via a virtual port 161 accessible through its vNIC (virtual Network Interface Card). This would be possible by implementing SNMP polling through the hypervisor. Any physical network device that can be virtualised has the support of the hypervisor to create virtual instances and manage the ports associated with them. Hence, in essence, it should be possible to equip each virtual device with an SNMP agent that maintains its own MIB (or a virtual MIB). Furthermore, the physical device itself will have an MIB making it a managed entity. The bottom line is that a network manager is able to monitor physical and virtual devices, albeit with intervention from hypervisor. Many variants of SNMP-based cloud management have been proposed; some of these are discussed in the next section.

5   SNMP-Based Cloud Management Schemes

In the recent years, there has been research towards providing network managers a view of the overall network where even virtual devices could be managed just as physical devices. It is important to consider that there are two dimensions for monitoring in a cloud. These are high-level monitoring which refers to monitoring of the virtual entities, and low-level monitoring pertaining to the hardware. Therefore, many schemes propose creation of management platforms that allows network managers to monitor and control application and device performance in all forms of virtualisation. This is in accordance with the requirement that IaaS infrastructures must be equipped with unified management platform that encompasses both physical and virtual networks [18]. Since the standard MIBs from RFC1213 does not provide everything needed to manage virtual devices, network management models for clouds rely on custom SNMP MIBs. In the studies described below, it can be seen that researchers mostly had to build newer MIBs or extend current ones to adapt to virtualised environments.

Currently, work is being carried out at Leeds Beckett University on using SNMP for developing schemes for sustainability in data centres. Although this research does not mirror network management in the traditional sense of diagnostics and performance management, it has weight in terms of traffic characterisation of virtual machines using SNMP. As a part of this research, Voderhobli [19] proposed a scheme to study traffic patterns and its impact on power. Due to the fact that there is a link between network traffic and power consumption, the research proposed to build an autonomic system to detect traffic characteristics in VMs. This model has three main elements in the overall system:

- VM Active Stats Poll Engine (VASPE): This is responsible for sending periodic SNMP queries to virtual machines on designated virtual ports. Each of the VMs has an agent and an MIB. Counters/values relating to all network related activities are recorded in the local MIBs, which is passed on to the VASPE when queried.
- Statistical Database: The VASPE is responsible for synthesising the data and storing it in the statistical database.

- Analytical Engine: The statistical database feeds into the function of the analytical engine. The role of the analytical engine is to analyse the statistics and initiate actions based on the learning. For example, the analytical engine could identify VM candidates for migration to other physical machines. The analytical engine also interacts with the substrate.

It is proposed that all of the above three elements are built into the hypervisor. A key feature of the above proposal is that it is closely in-line with the MAPE-K model where any action taken is heavily reliant on the knowledge base. In this case, the knowledge base is in the form of a statistical database. The work is still in its infancy as initial models are being built. In addition, the focus is purely on SNMP for green communications rather than a fully fledged network management for virtualised devices.

In an interesting approach to managing VMs, Peng and Chen [20] created a SNMP-based management system that relied on two agents. The default SNMP agent resided on the host machine whilst an extensible agent (AgentX) was used to collect live statistics from VMs. In order to avoid SNMP port 161 collision, AgentX was made to work on a different port. The Agent on the physical host was used to get values from the MIB that related to the overall statistics of the host. AgentX on the other hand was used to get statistics at the granularity of each VM. The authors refer to this as "dual SNMP agent scheme". The MIB used by AgentX is a custom enterprise MIB that includes objects specific to Virtual Machines. Some of the objects used in the MIB are:

vmDynamicIndex: for each spawned VM, this object maintains a unique index.
vmDomName: Name of the VM
vmDomMemoryUsage: Memory usage of the VM
vmDomVCPU: CPU cores allocated for the VM.

The above paper presented the working of the management system, which showed statistics on a per-host and per-VM basis.

In a research carried out by Hillbrecht and de Bona [21], the researchers created a management station capable of querying a set of VMs hosted on a physical machine. The objects in this MIB (which was called virtual-machines-MIB) recorded a range of objects related CPU usage, storage, disk images and network connections. They maintained tables to record creation of new VMs and further to change their configuration information. The management station sent SNMP queries to Virtual Machines. A listener SNMP agent on the virtual machines responds to queries sent by the management station. One of the interesting features of this management system is that it was not just a monitoring platform, but it also allowed for a network manager to exercise some control on VM operations. For example, a VM could be deleted by a network manager by writing into a MIB table that maintained the status of virtual machines.

The pieces of research briefed above were specifically aimed at management of virtual machines. However, applying SNMP to Network Virtualisation can prove to be significantly complex and will most certainly require more MIB objects to gather live statistics from virtualised network devices, as the following examples show.

Very recently, Asai et al. [22] proposed a Management Information Base for virtual machines controlled by a hypervisor. This is called VMM-MIB; it contains objects related to the hypervisor, resources, network access and storage. The complexity of these custom MIBs can be appreciated by looking at an excerpt from the MIB below (full draft available from IETF):

```
--vmMIB
            +--vmNotifications
                        +--vmRunning
                        +--vmShuttingdown
                        +--vmShutdown
                        +--vmPaused
                        +--vmSuspending
                        +--vmSuspended
                        +--vmResuming
                        +--vmMigrating
                        +--vmCrashed
                        +--vmBlocked
                        .
                        .
                        .
                        .
            +--vmObjects
                        +--vmHypervisor
                        ---vmNumber
                        ---vmTableLastChange
                        +--vmTable
                        +--vmCpuTable
                        +--vmCpuAffinityTable
                        +--vmStorageTable
                        +--vmNetworkTable
```

Da Paz Ferraz Santos et al. [23] argue that although the above MIB is able to represent software defined virtual routers (VR), the MIBs do not offer means of controlling VR due to the MIB having predominantly read-only objects.

Daitix et al. [24] used SNMP management in context of VRs. The authors acknowledge that due to its standardisation and simplicity, SNMP becomes a natural choice for managing virtual devices. They created a reference model to manage the physical host and several VRs. Two approaches were considered—the first, through a single SNMP agent placed in the substrate of the physical router. Their argument on having a single agent in the substrate is that it allows for management of physical host and the virtual instances. SNMP queries are tagged with SNMP community strings that offer distinction between the different VRs and the physical substrate. The second approach is to have independent SNMP agents for the physical router and the virtual router. SNMP queries are directed to network addresses of the VRs as each instance will have their own IP address. In either case, SNMP messages are relayed through the hypervisor, which is supported by APIs (Application Programming Interface) to handle SNMP communications. A custom

virtual router MIB was created inheriting and extension of a previous implementation of a similar MIB defined by Stelzer et al. [25]. This was done to alleviate the
shortcomings of standard MIBs in representing M:1 mappings. The authors mention as an example that bindings of several virtual ports to physical ports cannot be
accommodated using standard MIB. Hence, they incorporated MIB objects that
record virtual interfaces associated with each VR instance. The literature associated
with their research also provides examples of sample operations like VR creation
and VR deletion by writing into MIBs (using setRequest) and allowing changes to
be reflected in real time.

6 The Future of SNMP in Cloud Computing

One of the reasons the area of network management is so diverse is because it is
often difficult to create a general management solution that would suit all types of
networks. This problem even existed long before the advent of cloud computing.
With such complex and distributed architectures, it is common for network managers to prefer bespoke management solutions rather than one off-the-shelf. The
various research endeavours described in the previous section all point towards this
fact. Even with proprietary management tools, it can often be found that the option
of using SNMP-based network data collection is included. For good or for worse,
SNMP has become the de facto standard for network management irrespective of
network design or scale. Hence SNMP in the realm of cloud computing is here to
stay. However, the efficiency of SNMP for virtualisation management can be
questioned. Although many working models of using SNMP in cloud management
have been demonstrated successfully, it does not necessarily mean SNMP is the
most efficient for the task. It is a well-known fact that SNMP is a very "chatty"
protocol. Whilst management accuracy is dependent on frequent polling of devices,
thought needs to be given to how much of a detrimental effect it can have on
network utilisation. Ironically, there have been cases where network management
traffic has contributed to poor QoS. In a cloud-based environment, such episodes
could be catastrophic. Hence, whilst the networking community continues to adapt
to new developments using older protocols (like SNMP), they are likely to be
implemented as part of larger systems capable of offsetting the inefficiencies of
older protocols.

7 Conclusions

This chapter described some of the most recent work carried out in applying SNMP
for virtualisation monitoring and management. Some of these may not be complete
solutions to virtualisation management. Nevertheless they are an indication of
SNMPs gaining popularity in context of cloud computing. The discussion pointed

to various customised implementations of SNMP-based management. It seems evident that the area of network management in general is playing "catch-up" with rapid developments in cloud computing. SNMP has done well in this regard. It is incredible that a protocol that was founded a few decades ago is being applied in a world of grid and cloud computing. However, there are challenges ahead as technologies evolve. For example, it remains uncertain as to how efficient SNMP would be in more advanced scenarios such as software defined networking (SDN) and network function virtualisation (NFV).

References

1. Kasemsap, K. (2015). The role of cloud computing adoption in global business. Delivery and Adoption of Cloud Computing Services in Contemporary Organizations. IGI Global
2. Kerravala, Z. (2014). A guide to network virtualisation (white paper) [Internet]. Retrieved from https://www.avaya.com/en/documents/a-guide-to-network-virtualization.pdf.
3. IETF. (1991). RFC 1213 management information base [Internet]. Retrieved from http://www.ietf.org/rfc/rfc1213.txt.
4. Marston, S., Li, Z., Bandyopadhyay, S., Zhang, J., & Ghalsasi, A. (2011). Cloud computing—The business perspective. *Decision Support Systems, 51*(1), 176–189.
5. Telecommunications Engineering Center—TEC. (2016). White paper on network virtualisation [Internet]. Retrieved from http://tec.gov.in/pdf/Studypaper/White%20Paper%20on%20NetworkVirtualization.pdf.
6. Wen, H., Tiwary, P. K., & Le-Ngoc, T. (2013). *Network virtualization: Overview in wireless virtualization. Springer briefs in computer science.* Berlin: Springer.
7. Jain, R., & Paul, S. (2013). Network virtualization and software defined networking for cloud computing: A survey. *Communications Magazine IEEE, 51*(11), 24–31.
8. Doherty, J. (2016). *Scalability and performance for virtual appliances. SDN and NVF simplified.* London: Pearson.
9. Buyya, R., Calheiros, R.N., & Li, X. (2012). Autonomic cloud computing: Open challenges and architectural elements. In *2012 Third International Conference on IEEE Emerging Applications of Information Technology (EAIT)* (pp. 3–10).
10. Arista. (n.d). Impact of virtualization on cloud networking [Internet]. Retrieved from https://www.arista.com/assets/data/pdf/VirtualClouds_v2.pdf.
11. Pooja, K. (2013). Applications of green cloud computing in energy efficiency and environmental sustainability. *Second International Conference on Emerging Trends in Engineering.*
12. Josyula, J., Wakade, M., Lam, P., & Adams, T. (2010). Assurance management reference architecture for virtualized data centers in a cloud service offering [Internet]. Retrieved from https://www.cisco.com/en/US/technologies/collateral/tk869/tk769/white_paper_c11-575044.pdf.
13. Jailani, N., & Patel, A. (1998). FMS: A computer network fault management system based on the OSI standards. *Malaysian Journal of Computer Science, 11*(1).
14. Colman-Meixner, C., Develder, C., Tornatore, M., & Mukherjee, B. (2015). A survey on resiliency techniques. *IEEE Communications Surveys and Tutorials, 18*(3), 2244–2281.
15. Alwabel, A., Walters, R., & Wills, G. B. (2015). A resource allocation model for desktop clouds. Delivery and Adoption of Cloud Computing Services in Contemporary Organizations. IGI Global.
16. Huebscher, M. C., & McCann, J. A. (2008). A survey of autonomic computing—Degrees, models, and applications. *ACM Computing Surveys (CSUR), 40*(3), 1–28.

17. Kephart, J. O., & Chess, D. M. (2003). The vision of autonomic computing. *Computer, 36*(1), 41–50.
18. Ando, T., Shimokuni, O., & Asano, K. (2013). Network virtualization for large-scale data centers [Internet]. Retrieved from http://www.fujitsu.com/global/documents/about/resources/publications/fstj/archives/vol49-3/paper14.pdf.
19. Voderhobli, K. (2015). Achieving the green theme through the use of traffic characteristics in data centers, green information technology—A sustainable approach. Morgan Kaufmann.
20. Peng, Y. S., & Chen, Y. C. (2011). SNMP-based monitoring of heterogeneous virtual infrastructure in clouds. *Proceedings of APNOMS*.
21. Hillbrecht, R., de Bona, L. C. E. (2012). A SNMP-based virtual machines management interface. *Proceedings of IEEE Fifth International Conference on Utility and Cloud Computing UCC-2012* (pp. 279–286).
22. Asai, H., MacFaden, M., Schoenwaelder, J., Sekiya, Y., Shima, K., Tsou, T., et al. (2014). Management information base for virtual machines controlled by a hypervisor, draftietf-opsawg-vmm-mib-01 [Internet]. Retrieved from http://tools.ietf.org/html/draft-ietf-opsawg-vmm-mib-01.
23. da Paz Ferraz Santos, P., Pereira Esteves, R., & Zambenedetti Granville, L. (2015). Evaluating SNMP, NETCONF, and RESTful web services for router virtualization management. *IFIP/IEEE International Symposium on Integrated Network Management (IM)*.
24. Daitx, F., Esteves, R. P., & Granville, L. Z. (2011). On the use of SNMP as a management interface for virtual networks. *Proceedings of the IFIP/IEEE International Symposium on Integrated Network Management*.
25. Stelzer, E., Hancock, S., Schliesser, B., & Laria, J. (2003). Virtual router management information base using smiv2, draft-ietf-ppvpn-vr-mib-05 [Internet]. Retrieved September 5, 2016, from https://tools.ietf.org/html/draft-ietf-ppvpn-vr-mib-05.

Part III
Cloud Services, Big Data Analytics and Business Process Modelling

Strategic Approaches to Cloud Computing

Dilshad Sarwar

Abstract Cloud-based services provide a number of benefits in areas such as scalability, flexibility, availability and productivity for any organisation. These benefits can be further enriched by considering opportunities, which will allow for organisational enrichment at a strategic level. It is important to align businesses, ICT and security strategies to enable more beneficial outputs overall. Moving to cloud computing needs to consider strategic objectives of businesses essentially how IT cost impacts on the needs of future developments for businesses and IT departments within large organisations. Strategically implementing a cloud strategy can also be considered as a disruptive technology. Disruptive technology can be considered favourable in terms of offering organisational benefits. The most immediate benefits are consistent with reducing cost technology ownership, communication time therefore increasing the time benefits thus allowing organisations to become more productive.

1 Introduction

Business performance can be boosted significantly if strategically there is a consistent, effective and clear cloud strategy. Profitability can be enhanced significantly as a result of a successful implementation. Business systems are prone to working around organisational functionality and employee work. In a means to change and enhance the IT infrastructure from a slow and difficult environment to one which is effectively efficient and productive [1]. To reap the benefits of a successful implementation requires strategic measures such as: A data management review for the organisation which is holistic. The changes required here would encompass integrating, improving and migrating the data under the quality of data, which goes across the organisation [2]. There is a requirement which needs to focus on one

D. Sarwar (✉)
School of Computing, Creative Technologies & Engineering,
Leeds Beckett University, Leeds LS6 3QR, UK
e-mail: d.sarwar@leedsbeckett.ac.uk

© Springer International Publishing AG 2017
A. Hosseinian-Far et al. (eds.), *Strategic Engineering for Cloud Computing and Big Data Analytics*, DOI 10.1007/978-3-319-52491-7_9

single repositioning which contains data which is consistent with high quality that will enable on demand analysing [3]. Improving and optimising business processes, in essence requires a number of tools which will allow for a more managed and improved business process environment [4]. The benefits which cloud computing allow for are enormous in any organisational format (IBM).

The move towards cloud does involve the push towards a more centralised and automated approach within the organisation [5]. In particular, consideration needs to be given to how the people and processes will adapt to the technology [1]. Gathering technical requirements are essential when the migration towards cloud is implemented. Strategically the organisations business goals need to be considered [3, 5]. There needs to be an analysis in terms of the technical requirements of staff (IBM). The software applications need to be scalable. Careful consideration needs to be applied to load balancing capabilities and not simply a generic implementation of events [5]. With a strategic analysis there is a need to redistribute resources in order to accommodate the strengths and weaknesses of the cloud [3]. Security considerations need to also be applied to in order to work with the organisational data centres [2]. Some questions for consideration include whether or not there is a requirement for services to run [1].

While the focus of IT application strategy enables organisations to have a key business strategy it is important to consider the benefits when embracing cloud within any enterprise [5]. Cloud offers a number of embracing opportunities, which require a structured methodology to embrace cloud services with the information technology format. If organisations work towards creating a holistic cloud strategy, this helps in terms of meeting organisational goals. Some of the key strategic benefits are highlighted below:

- It increases the benefits which businesses can utilise within cloud. The advantages of this approach are phenomenal in terms of ensuring that the costs are reduced and efficiency is increased.
- The hidden benefits and the business benefits become clearer. The method used to create the cloud strategy can bring about opportunities which the organisations had failed to realise.

One of the fundamental benefits an organisation can succeed by, implementing cloud, is that cloud allows organisations to develop innovative ideas in line with organisational practices [4]. By adopting certain aspects of the organisation to cloud will increase the speed of building, testing and refining new applications, this enables uses to access and manage new ideas in order to understand what practices work effectively and what practices are unable to work effectively within the organisation. If the strategy for cloud is thought through this, it enables a greater opportunity. If the cloud requirements are addressed in terms of a shared, virtual infrastructure of storage, the network resources can become more efficiently in relation to the adopted cloud model [4]. In today's IT demanding competitive environments regardless of ability the access to cloud is significant, this is essentially what organisations should consider holistically when considering implementation.

2 Effective Organisational Practices When Considering Cloud Strategy

Enabling and determining the workloads moving towards the cloud model, with the view to prioritise the move of workload migration. These decisions can be deemed quite complex [6]. Complexity arises as there are a number of variables, which need to be considered—ranging from the direction the workload is implemented, the operational flexibility, the lifecycle and the goals associated with the categorisation [7]. The two fundamental areas of strategy embraced within cloud relate to the strategic constructs indicating the level of intensity within the workload context [8]. The operational flexibility, which looks at the infrastructure of the applications and how key it is in terms of the resources in place for organisations to benefit from an effective cloud strategy [8]. An example of this implies that the custom application is encompassed within an array of numerous applications but has limited requirements in terms of resource flexibility [9].

The strategic value of cloud focuses on operational flexibility—looks at addressing the changing business needs and changing to the variable workload. Strategic Value—What values are fundamental to the business and what drives the business differentiation. It is important to determine which type of cloud constructs best fit the requirements of the workloads [10]. There are numerous categories that are appropriate for cloud concentrating on workload tasks. The following are the best-categorised cloud types:

Private cloud ensures that this type of cloud can be developed and manipulated by internal IT or service providers external to the organisation based on the location, within the organisational firewall [6]. These services are then presented to the organisations internal users through a service category, which allows user access to enable users to have access to the variety of capabilities [11]. Workloads and applications used, which are mostly associated with private cloud, are more inclined to have a high level of strategic importance and increased level of operational requirements of flexibility [12]. This can be an area, which is fundamental to business requirements which allows for differentiation to occur and hold a degree of sensitive and confidential information [3]. The typical examples here require a very high level of flexibility as the resources used need to be focused on being high level in terms of the organisation. Public cloud ensures that these services which are associated and implemented by a third-party service provider. Hybrid cloud provides the ability to associate private cloud with the public cloud which focusses on a pay for what you use model.

The cloud type which is the most appropriate cloud model for the organisation: public cloud enables the use of applications related to the level of workloads for the use of public cloud and have low overall value and high level of operational movement. These may not be fundamental to the organisation and need to move towards demands which are technical and enable flexible requirements [3]. Additionally other flexible requirements such as a high level of usage which can be taken up by an increased level of usage, the pay-as-you-go model is adapted with ease [3].

Follows a working environment which are low/high strategic consideration and therefore good candidates for using this type of cloud environment [13]. When the correct type of workloads can be defined, then the most appropriate method can be applied [14].

With cloud systems, there are considerable benefits outlined by the cloud and the implementation aspects of it for cloud-based environments [15]. These benefits are outlined by key areas such as the cloud environment creating greater flexibility and ensuring that there is a greater degree of emphasis on scalability, the level of availability and the overall productivity. The cloud enables a greater degree of opportunities as argued [16], risks and new forms of development [12]. One must endeavor to consider the delivery models and create a focus which will be aimed at developing a lower level of risk, thus ensuring the computer structures are clearly implemented within the sustainability concepts of business [14].

The overall time spent in implementing cloud ensures that there is a good overall plan in place which will allow for the successful implementation of cloud [6]. The impact on the organisation holistically in terms of developing new opportunities requires the need to address that the level of shared resources are implemented within the strategic direction of the organisation [17]. The organisation furthermore needs to focus on ensuring there are capabilities which the organisation is able to clearly consider as outsourcing [18]. The services that are implemented by cloud do need to think about considering specialisms in terms of enterprising the overall structure and architecture [19]. It is imperative to understand that the cloud solutions cannot impact on the overall organisational processes [19].

Managing organisational change through the implementation of cloud requires concerns raised to be addressed in a form of structured and analytical processes [13]. The governance aspect of implementing cloud essentially requires a significant amount of support and guidance in order to ensure that the infrastructure outlined for the organisation is introduced and applied correctly [20]. The notion of implementing cloud requires essential consideration in terms of ensuring an effective outsourcing solution for the organisation [12]. Therefore a full structured analysis of the organisational processes need to be considered before any form of implementation is considered [21]. The business model adopted needs to also consider applying a structured methodology in line with the organisation requirements [21].

Organisational risks need to be addressed and applied in terms of ensuring that the cloud strategy adopted is in line with the business models [7]. Risk management and risk analysis are essential aspects for consideration before implementation occurs [22]. A number of risks need to be considered before implementing cloud computing. These risks include the functionality of the model which has to be implemented within the organisation [22]; the functionality costs which are required to implement a structured and workable cloud computing environment. There should be some fundamental consideration of applying accurate and beneficial functionality within the cloud infrastructure [11]. Standards and performance requirements are required to be applied to the process of implementation to ensure the smooth and seamless implementation and migration to cloud [13]. The assurance is that the cloud environment can be utilised effectively. Other concerns that need to be applied include security and the actual compliance of the legislation and obligatory considerations are applied in terms of the regulations which need to be adhered to Zhang et al. [8]. Additionally, requirements need to be applied to the overall business organisation [21]. The benefits and the overall risks need to be considered. The organisation wishing to introduce and migrate towards cloud are required to ensure that there is a clear business case for a successful cloud migration [23].

To be able to manipulate data effectively and efficiently, the organisation needs to believe that migrating towards cloud will in effect allow it to achieve the greatest of benefits [8]. As highlighted in previous sections, it is imperative that the infrastructure is strong, robust and workable [1]. To build a strong cloud strategy, a number of fundamental aspects need to be considered [24]. Cloud integration does require strong and holistic management for migration towards cloud to be effective [22]. A clear understanding of the real organisation environment is required at all levels [19].

Systems that are already in existence with organisations create the most difficult barriers of allowing for the implementation of cloud technology [1]. Strategically considering organisational cloud migration involves a number of cost issues which have a huge impact on the financial elements within the organisation [10]. Cloud computing has the means to become a major effective element within an organisation with the potential to create an effective environment and positively impact on the work environments and demands [1]. A legacy system which is already in place is required to be monitored and updated which would in essence impact on the cloud infrastructure [3]. Considering cloud migration does require changes to the data centres infrastructure—the strategic considerations should allow for the data centre infrastructure to address the change requirements and also prepare the organisation to move towards cloud computing [1]. The organisation needs to assess the impact of the data centre infrastructure and assess the impact this has on the IT teams within the organisation [5].

The means of moving towards migration involves the requirements to migrate creating a broader scope of disaster management of data [4]. It is essential if not critical to gain a clear sense of developing and implementing a business case [2]. The basics for implementing cloud may also include the need to increase the level of workload, but needs to consider the level of work load and limitations associated with each department [8]. Working towards implementing a cloud computing environment, the existing infrastructure does require full consideration (IBM) [25]. It is essential for all levels of the organisation to understand that all IT teams which include the software and hardware require consideration [1]. The IT group including the network and storage administrators need to establish and consider these points [26]. The information provided by these groups is essential when making informed decisions about changes to the strategic changes of systems [27].

Assessing the existing infrastructure is essential. When addressing the requirements to move towards a cloud model, the existing structure needs to be analysed [5]. The strategic need to ensure migration for cloud bridges the gap between the public cloud [1]. Public cloud preparation requires a number of cloud portals which can be connected/linked on the premises [8]. There are a number of areas to consider here—it is important to understand or gain an insight as to how the existing infrastructure can combine the move towards cloud migration [5]. It is important to acknowledge whether or not a completely new infrastructure is required or will it be feasible to continue with the existing infrastructure can combine the migration towards cloud [11]. Consideration should be applied to the use of other tools such as VMWare v Cloud Connector, which would allow for immediate migration to cloud [12]. Additional considerations maybe that the tools can allow for access and providers at different locations [2]. Can these tools allow for chargeback and real-time reporting of costs and performance metrics, is the new Service Level Agreement (SLAs) being adhered to? There needs to be consideration in terms of Service Level Agreement. How are templates and configurations going to be managed [3]. The question of the management of authentication needs to be considered. Is there going to be an audit trail? These questions are imperative when considering migration to cloud [8]. Public cloud benefits need to be looked at —this is the ability of public cloud which allows for resourcing to match the work load, this becomes very cost effective in the long run; furthermore there is a no requirement to consider a yearly peak workload, requires just a day-to-day work load basis [5].

There is a requirement that staff needs to consider and maintain good virtual machine templates which are required to use the tools that are necessary for cloud migration [1]. The strategic approach to migrating towards cloud attempts to consider the change control that ultimately prevents outage and supports all OS configurations ensuring all OS configurations are synchronised. Staff training is also essential when considering the migration towards cloud [8]. Additional infrastructure requires consideration; firewall rules need to be adhered to even more so than previously [2]. Networking is essential for cloud possibilities [3]. Success can be achieved with migration of cloud implementation which essentially is dependent on good networking practices [12]. Moving workloads externally to a public cloud has to adhere to network connection redundancy [4].

3   Security

This is an essential requirement that needs to be considered [28]. There needs to be some consideration over how the tools and the cloud provider interact with the data centre and grant them access through network and host-based firewalls if necessary [19]. This may be difficult as private cloud management interfaces are on completely internal, private networks [4]. Cloud infrastructure impacts include implementation of authentication and access control for the new hybrid cloud should be considered [18]. Access control policies in its user database are required. There are some additional considerations that need to be noted [5]. Additional requirements involve knowing how much network and storage I/O that your applications generate enables you to develop connections [9] and strategic consideration such as analysing technical requirements and applying protocols. With the best planning there can still be huge obstacles [5]. When preparation has been put into place for cloud migration there needs to be some considerations provided to the data centre configuration management, the network status and storage Armbrust et al. [1]. Cloud migration does have a tremendous impact on the existing data centre infrastructure [14]. The overall consensus illustrates that cloud migration allows for a more flexible and a more advantageous solution for the overall organisations strategic capability [29].

The data centre infrastructure for cloud migration needs to be prepared [5]. Consideration needs to be given to the application and that the system administrators liaise with each other which will allow network engineers to consider sizing and troubleshooting [2], as a result of the move towards or migrating towards cloud [24]. Moreover, intrusion detection and prevention systems are required to be considered in order that communication from remote hosts are not interrupted [3]. Strategic consideration—strong monitoring technology is required to state and perform with the data centre [13]. Moreover, intrusion detection and prevention systems are required to be considered in order that communication from remote hosts are not interrupted [5].

Strategic consideration requiring strong monitoring technology is required to state and that hosts are not interrupted [30]. Strategic consideration—strong monitoring technology is required to state and performance with the data centre [4]. As there is a move towards the cloud highlighted in terms of whether these systems are extensible [31]. Disaster recovery needs consideration in order to understand if the primary site is down the consideration to how the system can be managed needs to be determined [8]. Strategic considerations in terms of whether or not there is a need to consider secondary monitoring system at the alternative site [18]. Importance should be applied to real-time metrics. Secondary monitoring at the alternative site needs to be evaluated [5].

Real-time performance metrics need to be considered in terms of the positive aspects of performance metrics which allow for technical staff to troubleshoot using tools that monitor service which can then automatically enhance an organisation's capability [13]. Good programming interfaces and having staff who can understand and manage tools that integrate them into the organisations business processes [5]. Adaptation between systems is essential; some adaptations are simply processes orientated rather than technological, although there are likely good integration possibilities for adaptation to be implemented successfully [13].

Cloud migration has an impact on existing data centre infrastructures which essentially require the need to consider the key processes involved in the migration towards cloud [13]. The infrastructure within any organisation requires a holistic overhaul. It is imperative not to ignore storage and backup processes [5]. Strategic considerations should also involve processes as it is essential that communication is flowing in terms of the business requirements and essential technical requirements [8]. Strategic considerations highlight that not all cloud storage and applications are the same. Cost considerations are essential when performance is required [24].

However, if the system in place requires added information [2]. Provider details are necessary requirements, initial considerations include discussion in terms of whether the legacy system will be used for backup processes [5]. How will network traffic be affected as a result of the move towards cloud? It might be that the cloud provider offers backup solutions internally that are more cost effective but require different processes and procedures that restore data than those already established systems [4]. Encryption of the system that is used need to consider third-party shared services [19]. Procedural changes are necessary when securing and storing encryption keys [2]. Moving workloads to cloud has a number of processes which are tarnished with pitfalls and potential problems [12]. The issues and problems aligned with the movement from a physical to a hybrid environment can be quite complicated which involve a number of technical problems to logistical and financial complications [13]. The movement of this essentially needs considerable detailed planning [5].

4 Conclusions

The strategy development that is required to be considered for successful migration include a number of areas which should be adhered to and the timelines applied that are required to be followed in terms of a means of upgrading to a new operating system or updating old or unsupported applications [4, 13]. Basic planning is necessary, grasping a need to create a cloud environment and then performing the migration after which any problems can be rectified [8]. Although there are a number of organisations that are happy to engage in an extensive review to look at and address the concerns of performance and compatibility concerns and identifying changes that are needed to be made prior to migration, the problem here is that these processes and procedures are a necessity to be followed [2].

Automation maybe possible if a migration is like to like and this inevitably will cut costs significantly [3]. It is important to understand that there is never any migration which can become hundred percent automated [8]. There is a concern over using automated migration tools for which there is a great deal of software installation, reconfiguring images, testing/troubleshooting and data syncing [5]. As a result, organisational needs are required to be evaluated in order to be more efficient and cost effective and implanting a migration strategy [23].

As organisations migrate and change every two to four years, clients are more than likely to move towards a new formulated operating system which is part of their migration process [5]. Older operating systems are more likely to have a compliance, security and support concerns as it is not a good idea to express sensitive and proprietary information unnecessarily—therefore upgrading makes more sense [24]. Movement towards establishing compatibility work is required in order that applications are functioning correctly in the new environment [5]. Application updates should be performed and tested ahead of any migration, coding or compatibility issues need to be addressed after the movement towards a cloud environment [5]. Understanding the overall timings and variables of the system does require consideration of a number of variables, which include basic information which encompasses the volume of data and the number of servers which are involved in the move and which carry out very significant move towards data synchronisation [14]. Resources are needed to be extensively prepared [5]. It is important to understand that existing staff should be focused on ensuring cloud migration processes are flowed to address and the code incompatibilities that require addressing [1]. There are also very few companies who are actually equipped to programmatic problems [5]. If the correct resources are not in place then there is a clear delay in place to make a professional fix [11]. It is important to address any concerns in terms of costings and clients [5]. Essentially developing a clear-cut migration strategy will ensure that successful cloud migration can be achieved [3]. Fundamentally it is important to prioritise the workload when moving towards cloud. There is concern related to the level of security of the date and the loss of administrative control [6]. Moving the workload to private cloud helps those organisations which have moved towards the utilisation of the virtual environment [19]. Ensuring priority is given to less critical smaller workloads to enable them to move towards the private cloud enables experience to be gained throughout the organisation [7]. Using trusted cloud services helps to formulate appropriate cloud environments [6].

It is important to understand that the public sector has increased in growth over the years [11]. Therefore it has become imperative that the ICT infrastructure is developed to match this established growth [6]. Any organisation wishing to migrate towards cloud needs to ensure that there is an ongoing development area [19]. The cloud approach clearly outlines cost savings with the increased level of flexibility [23]. Furthermore, the move towards cloud migration reduces the level of bureaucracy the cost implications and the overall management of the systems [6].

References

1. Armbrust, M., Fox, A., & Griffith, R. (2010). A view of cloud computing. *Communication of the ACM*, 50–59.
2. Bera, M., Sudip, J., & Rodrigues, J. P. (2015). Cloud computing applications for SMART grid a survey. *IEEE Transactions on Parallel and Distributed Systems, 26*(5), 1477–1494.
3. Bhat, J. M. (2013). Adoption of cloud computing by SMEs in India: A study of the institutional factors. *Nineteenth Americas Conference on information Systems* (pp. 1–8).
4. Folkerts, E., Alexander, K., Sacks, A., Saks, A., Mark, L., & Tosun, C. (2012). Benchmarking in the cloud: What is should, can and cannot be.
5. Choudhury, V. (2007). Comparison of software quality under perpetual licensing and software as a service. *Journal of MIS, 24*(2), 141–165.
6. Islam, T., Manivannam, D., & Zeadally, S. (2016). A classification and characterisation of security threats in cloud computing. *International Journal of Next Generation Computing*.
7. Floerecke, S., & Lehner, F. (2016). Cloud computing ecosystem model: Refinement and evaluation.

8. Zhang, Y., Li, B., Huang, Z., Wang, J., & Zhu, J. (2015). SGAM: Strategy—Proof group buying—Based auction mechanism for virtual machine allocation in clouds. *Concurrency and Computing: Practice and Experience, 27*(18), 5577–5589.

9. Dasgupta, K., Mandal, B., Dutta, P., Mandal, J. K., & Dam, S. (2013). A genetic algorithm based load balancing strategy for cloud computing. *Procedia Technology*, 340–347.

10. Li, J., Chen, X., Jia, C., & Lou, W. (2015). Identity based encryption with outsourced revocation in cloud computing. *IEEE Transaction on Computers*, 425–437.

11. Yang, C., Xu, Y., & Nebert, D. (2013). Redefining the possibility of digital earth and geosciences with spatial cloud computing. *International Journal of Digital Earth*, 297–312.

12. Ward, J., & Peppard, J. (2016). The strategic management of information systems: Building a digital strategy.

13. Krebs, R., Momm, C., & Krounev, S. (2012). Metrics and techniques for quantifying performance isolation in cloud environments. ACM QoSA.

14. Rittinghouse, J. W., & Ransome, J. F. (2016). Cloud computing: Implementation, management and security.

15. Chang, J. F. (2016). Business process management systems: Strategy and implementation.

16. Grant, R. M. (2016). *Contemporary strategy analysis.* Wiley.

17. Chen, M., Zhang, Y., Li, Y., Mao, S., & Leung, V. C. (2016). Understanding behavioural intention to use a cloud computing classroom: A multiple model comparison approach. *Information and Management, 53*(3), 355–365.

18. Zhan, Z. H., Liu, X. F., Gong, Y. J., Zhang, J., Chung, H. H., & Li, Y. (2015). Cloud computing resource scheduling and a survey of its evolutionary approaches. *ACM Computing Surveys*, 63.

19. Sookhak, M., Gani, A., Talebian, H., Akhunzada, A., Khan, S. U., Buyya, R., et al. (2015). Remote data auditing in cloud computing environments: A survey, taxonomy, and open issues. *ACM Computing Surveys (CSUR), 47*(4), 65.

20. Zhou, B., Dastjerdi, A. V., Calheiros, R. N., Srirama, S. N., & Buyya, R. (2015). A context sensitive offloading scheme for mobile cloud computing service. *International Conference on Cloud Computing IEEE* (pp. 869–876). IEEE.

21. Li, W., Zhao, Y., Lu, S., & Chen, D. (2015). Mechanisms and challenges on mobility-augmented service provisioning for mobile cloud computing *53*(3), 89–97

22. Puthal, D., Sahoo, B. S., Mishra, S., & Swain, S. (2015). Privacy-aware adaptive data encryption strategy of big data in cloud computing. *Computational Intelligence and Networks*, 116–123.

23. Zhang, H., Jiang, H., Li, B., Liu, F., Vasilakos, A. V., & Liu, J. (2016). A framework for truthful online auctions in cloud computing with heterogeneous user demands. *IEEE Transactions on Computers, 65*(3), 805–818.

24. Hashem, I. T., Yaqoob, I., Anuar, N. B., Mokhtar, S., Gani, A., & Khan, S. U. (2015). The rise of 'big data' on cloud computing: Review and open research issues. *Information Systems*, 98–115.

25. IBM. (n.d.). What is cloud computing? Retrieved from http://www.ibm.com/cloud-computing/us/en/.

26. Gai, K., Qiu, M., Zhao, H., & Xiong, J. (2016). Privacy-aware adaptive data encryption strategy of big data in cloud computing. *International Conference on Cyber Security and Cloud Computing*, 273–278.

27. Coutinho, E. F., Carvelho, d., Rego, F. R., & Gomes, D. G. (2016). Elasticity in cloud computing. *IEEE 3rd International Conference on Cyber Security and Cloud Computing* (pp. 273–278).

28. Gkatzikis, L., & Koutsopoulos, I. (2013). Migrate or not? Exploiting dynamic task migration in mobile cloud computing systems. *IEEE Wireless Communications*, 24–32.

29. Assuncao, M. D., Calheiros, R. N., Bianchi, S., Netto, M. A., & Buyya, R. (2015). Big data computing and clouds: Trends and future directions. *Journal of Parallel and Distributed Computing*, 3–15.
30. Chen, X. (2015). Decentralised computation offloading game for mobile cloud computing. *IEEE Transactions on Parallel and Distributed Systems*, 974–983.
31. Alkhanak, E. N., Lee, S. P., & Khan, S. R. (2015). Cost-aware challenges for workflow scheduling approaches in cloud computing environments: Taxonomy and opportunities. *Future Generation Computer Systems*, 3–21.

Cloud Security: A Security Management Perspective

Mohammed M. Alani

Abstract This chapter discusses, on strategic level, security considerations related to moving to the cloud. It starts by discussing common cloud threats with brief explanation of each of them. The next section discusses the detailed differences between the cloud and classical data-center in terms of security. Presented in this chapter, the process of prioritizing assets and threats when moving to the cloud. The chapter explains how to evaluate the risks on each asset and how to prioritize the required spending on the information security budget. The chapter concludes with a discussion of general security consideration for the cloud.

1 Introduction

Every Chief Information Officer (CIO), Chief Information Security Officer (CISO), and even Chief Executive Officer (CEO) worries about the security of their data and the availability and continuity of their systems. When a major decision, like moving to a new paradigm, is to be made, a very thorough study needs to be carried out. The long-term and short-term effects of moving to the cloud must be studied well before the move. With many high performance public cloud service providers available in the market, the decision of which cloud provider to choose relies mostly on the security provided by this service provider.

Many managers have concerns regarding security of their data on the cloud. Some managers think that the benefits out weigh the risks and decide to take the leap to the cloud. These benefits can be briefly listed in the following points [5]:

1. Cost Saving
2. Scalability and Flexibility.
3. Reliability.
4. Reduced IT Technical overhead and Management Efforts.
5. Reduced Environmental Effect.

M.M. Alani (✉)
Al-Khawarizmi International College, Abu Dhabi, UAE
e-mail: m@alani.me

© Springer International Publishing AG 2017
A. Hosseinian-Far et al. (eds.), *Strategic Engineering for Cloud Computing and Big Data Analytics*, DOI 10.1007/978-3-319-52491-7_10

6. Better Hardware Resources Utilization.
7. Ease of capacity planning and increased organizational agility.

One of the main reasons for many organizations to move into the cloud is to reduce the costs. Running your own data center can be resource-exhausting even to large organizations.

In 2010, [22] introduced one of the earliest comprehensive overall security perspective of cloud computing with a highlight on a group of security concerns. This paper contained a useful start for people interested in moving to the cloud.

In 2011, [6] suggested a cloud security management framework that was the base of more research later on. The proposed framework was based on aligning the Federal Information Security Management Act (FISMA) standard to fit the cloud computing model. This helped cloud computing service providers and consumers to be security certified.

A more comprehensive and detailed introduction to cloud security and security management was recently introduced in [24]. The book discussed in details the security concerns and security management frameworks that can help nontechnical audience in gaining knowledge about the topic.

In the next section, we will discuss the most commonly known threats in cloud computing. Afterwards, we will point out the main differences between cloud computing and a classical data-center. The fourth section will discuss the numbers behind security controls employed in the cloud and how to prioritize the spending. The final section explains critical cloud security considerations.

2 Common Cloud Threats

A threat, as identified by Internet Engineering Task Force (IETF), is a potential for violation of security, which exists when there is a circumstance, capability, action, or event that could breach security and cause harm [26]. In their "The Notorious Nine" report, Cloud Security Alliance (CSA) have identified nine threats that represent most important threats to cloud computing security in the year 2013 [28]. These threats will be discussed in the coming subsections in order of importance as mentioned in [28].

2.1 Data Breaches

Big organizations spend a lot of money to keep their data safe from the prying eyes of a malicious attacker. Losing sensitive clients data can lead to severe business consequences that can go all the way to shutting down the whole business. This argument is valid even for small organizations. The owner of a small e-commerce website might not spend a lot on protecting clients' sensitive data like credit card numbers. However, the same website owner will be liable and subjected to legal consequences that would cost him or her much more than what the business is worth if this information is stolen.

It is definitely terrifying to consider a data breach as large as what happened to Sony in 2011 [12]. Terabytes of private data was stolen by attackers. Attackers also deleted the original copies from Sony's computers, and left messages threatening to release the information if Sony did not comply with the attackers demands. After the breach, 32,000 private internal and external emails were released in public. Passwords and personal information of actors, managers, and Sony staff were publicly available. This data breach's cost is still growing as you read this.

Lets look at a scenario where a multitenant cloud service is poorly designed. Research in [9] have shown that if a flaw exists in one clients application, it can lead to allowing the attacker to access the data of that client and all other clients hosted on the same physical machine.

Before we proceed in explaining how this threat might be realized, we need to define side-channel attacks. In a *side-channel attack*, the attacker gains information about the cryptographic technique currently in use through detailed analysis of physical characteristics of the cryptosystem's implementation [4]. The attacker uses information about the timing, power consumption, electromagnetic leaks,... etc. to exploit the system. This collected information can be employed in finding sensitive information about the cryptographic system in use. For example, information about power consumption can result in knowing the key used in encryption [32].

Researchers introduced a side-channel attack, in [30] that enables one virtual machine (VM) hosted on a physical machine in the cloud to extract private cryptographic keys used by another virtual machine hosted on the same physical machine. This attack is just an example of how poor design can cause severe data breaches.

The first solution that comes into your mind when we discuss data breach is definitely encryption. One way of mitigating data breach risk is to encrypt all of the clients data. However, this might not be as easy as you think, and we will discuss why. All encryption types are done with the help of a key. To keep information secure, the key should be with the client only and not stored on the cloud itself. However, if the encryption key is lost, the client would have a complete data loss. Thus, the client would need to have a backup copy of the data, somewhere else, or even offline backup. The client should keep in mind that having more copies of the data would potentially increase the probability of data breaches. Generally, encryption is not as straightforward as it is with data stored at your own servers. When employing a public cloud, the applicability of encryption is far from simple. Many questions will arise when we talk about encryption;

- should the encryption happen at the client side or cloud side?
- if at the client side, where should the client keep the key?
- if at the cloud side, should the encryption keys be stored at the service provider side or the client side?
- how can we safely transfer the keys to the cloud?

- if the encryption happens at the client side, where should decryption happen?
- what happens is the user loses the encryption keys?

and so many other questions that makes the whole process an exhausting burden. You can find more details on the challenges of encryption in [10, 27].

2.2 Data Loss

The threat of data loss has been there since the invention of computers. Many factors can take part in causing the data loss. The following, nonexclusive, list shows common reasons of data loss [5]:

1. Malicious attacks.
2. Natural catastrophes like earthquakes, floods, and fires.
3. Accidental erasure or loss by the cloud client organizations' staff.
4. Accidental erasure or loss by the cloud service provider.

Looking at this list, you can easily see that protection from data loss does not fall into the duties of the cloud service provider alone rather than the client organization as well. When the loss is caused by a malicious attack, most clients would blame the cloud service provider. However, this is not always the case. To elaborate more on this point, we will look at two example attacks that happened in 2011. In the first example, a cloud-based Nasdaq system named "Directors Desk" was targeted by attackers. The system aims to facilitate communication between around ten thousand senior executives and company directors. By having access to this system, attackers could eavesdrop on private conversations between executives that can be used as stock-market leaked information to benefit competitors. While attackers had not directly attacked trading servers, they were able to install malware on sensitive systems, which enabled them to spy on dozens of company directors [1].

Our second example that took place in 2011 was the Epsilon attack. Epsilon is a cloud-based email service provider that went under attack in April, 2011. The attack was a spear-phishing attack. While a *phishing* attack is a type of social-engineering attack in which attackers use spoofed email messages to trick victims into sharing sensitive information or installing malware on their computers, spear-phishing is a more complex type of phishing. Spear-phishing attacks use specific knowledge of individuals and their organizations to make the spoofed email look more legitimate and targeted to a specific person [16]. In the attack on Epsilon, data of 75 business organizations were beached and the list was growing. Although Epsilon did not disclose the names of companies affected by the attack, it is estimated that around 60 million customer emails were breached.

Looking into the details of the previous two examples, you will see that most of them happen because of some sort of misuse by the client. Attacks like spear-phishing, weak passwords,…etc. are mainly caused by the client rather than the service provider. At the end, we cannot come up with a general rule of "whose fault is it?" when a malicious attack happens. As mentioned in the list of common reasons of data loss, malicious attacks are not the only cause of concern. Natural catastrophes can be a serious cause of concern as well. They are not controllable by neither the client nor the service provider. On the other hand, clients and service providers can reduce the impact of these catastrophes by implementing proper counter measures.

Accidental erasure or loss by the client's staff can happen in multiple ways. An example of this case is the one mentioned in the previous section; if the client encrypts the data before uploading to the cloud and loses the encryption key, data would be lost. Client's data can also be erased or lost by the cloud service provider. This erasure can happen deliberately or accidentally. Either way, many organizations dislike the cloud because of this fact. Giving a service provider the control over your data storage is not something that many organizations feel comfortable doing. This is mainly the reason why many organizations create private clouds. In many countries, the organizations are required to keep complete audit logs of their work. If these logs were stored on the cloud and lost, this can jeopardize the existence of the organization and cause many legal issues.

Data loss is considered a threat to the IaaS, SaaS, and PaaS models. Mitigation of this threat can be done through backups. Regular (daily or even hourly) offline backups can be used to restore data with minimum loss. For services that have zero-tolerance for data loss, online backups with a different service provider can be a costly, but safe, solution.

2.3 Account or Service Hijacking

The risks caused by older attacks like social engineering, exploiting software vulnerabilities are still valid. These attacks can still achieve the intended result for a malicious attacker. Reusing of usernames and passwords magnifies the severity of this threat. In the previous subsection, we have discussed an example of cloud attacks based on the social engineering tricks, like spear-phishing and have shown its magnitude in attacks like the one on Epsilon, the cloud-based email service provider.

In cloud computing, this threat takes a new dimension. After gaining access to the clients credentials, attackers can eavesdrop on the client transactions, return falsified information, manipulate data, and even redirect the users to illegitimate sites. In addition to that, the attacker can use the instances of the client as attacking bases to attack other servers. Such access, can compromise confidentiality, availability, and integrity.

In 2009, Amazon had a large number of their cloud systems hijacked and were used to run Zeus botnet nodes [2]. Zeus is a banking trojan and one of its variant was spotted using the Amazon's cloud service as a command and control channel for infected machines. After the target gets tricked into installing the password-logging malware, their machine began reporting to EC2 for new instructions and updates. On their side, Amazon said that the trojan was using a legitimately bought service that had been compromised using some bug.

According to [3], in 2010 Amazon.com had a cross-site scripting (XSS) bug that allowed attackers to hijack credentials from the site. The bug on Amazon allowed attackers to steal the session IDs that are used to grant users access to their accounts after they enter their password. It exposed the credentials of customers who clicked on a specific link while logged in to the main Amazon.com page.

2.4   Insecure Interfaces and APIs

If the client want to manage and interact with the cloud services, the cloud service provider needs to provide a set of Application Programming Interfaces (APIs). These APIs are used for provisioning, management, orchestration, and monitoring. Availability and security of the cloud service is heavily dependent on the security of these APIs. Securing the system becomes more complex when the organization builds on these APIs to provide value-added services to their clients. This dependence on APIs shifts their architecture into a layered model. This layered model increases risk by increasing the exposure area of the system. In many scenarios, the organization will have to pass their credentials to a third party to enable them to create or use these new APIs.

It is essential that the cloud service clients understand the security implications that come with the usage, management, and monitoring of cloud services.

It is also essential to select a cloud service provider that provides authentication and access control, encryption, and activity monitoring APIs that are designed to protect against accidental as well as malicious attempts to circumvent the policy.

Poorly designed APIs can cause major threats to confidentiality, integrity, availability, and accountability. Thus, secure and properly designed APIs must be a vital part in the client's cloud service provider selection criteria.

2.5   Threats to Availability

Denial of Service (DoS), as a threat, exists in almost all networking services. In general, DoS is preventing the service from being provided to its intended audience. This can be through preventing website visitors from viewing the website, blocking legitimate user access to a Voice-over-IP (VoIP) server, ... etc. DoS in cloud computing would not only render the service unavailable, but cause huge additional financial implications. Since cloud service providers charge their clients based on the amount of resources they consume, the attacker can cause a huge increase in the bill even if the attacker did not succeed in taking the client's system completely down. Another point that makes this threat even more dangerous in cloud computing is that cloud computing clients share the same infrastructure. Hence, a heavy DoS attack on one client can bring down the whole cloud and affect other clients.

2.6   Malicious Insiders

According to [29], 62% of security professionals saw increase in insider attacks. In the same survey, 59 % of security professionals believe that privileged users like managers with access to secure data are most risky. Another part of the statistics show that 62% of security professional believe that insider attacks are very difficult to detect.

Despite the low probability of occurrence in comparison to external attackers, a malicious insider can cause a lot of harm cloud computing. In [31], it is considered one of the highest possible risks on a cloud computing service. The reason behind that is that cloud architectures necessitate certain roles which are considered of the highest possible risk. An example of these roles is CP system administrators and auditors and managed security service providers dealing with intrusion detection reports and incident response.

Organizations that depend solely on the service provider in security are at great risk due to malicious insiders. Strict policies must be applied on the client organization's side to reduce the risk of malicious insiders.

Encrypting the client data will not completely mitigate this threat. If the encryption keys are not stored with the client and are only available at data-usage time, the system is still vulnerable to malicious insider attack. Thus, it is advisable that all client data is encrypted and the keys should be kept with the client to reduce the risk of malicious insiders. However, this would raise other issues related to data loss and data breach.

2.7   Abuse of Cloud Service

The huge computing resource available to cloud service clients is also available to attackers and service abusers. The access to cloud computing resources can be abused and these resources can be directed towards attacking other systems.

The imminence and severity of this threat has dropped over the past few years due to stricter policies followed by cloud service providers.

The ease of deployment of on-demand processing power available in the cloud makes it a very tempting tool for attackers. The anonymity provided by IaaS and PaaS service models also unveil critical exploitation possibilities. This anonymity can lead to abuse of the provided infrastructure in conducting DDoS attacks, controlling an army of botnets, hosting malicious data, unlawful distribution of copyrighted data, and last but not least, sending spam emails.

2.8 Insufficient Due Diligence

Many organization aim at the quick adoption of cloud computing. Organizations that does not fully comprehend the requirements of proper implementation of cloud computing will have many issues with operational responsibility, incident response, and many other aspects [28].

Organizations with weak understanding of cloud computing can have contractual issues over obligations on liability, response, or transparency. This is caused by mismatch of expectations between the client and the cloud service provider. Security issues might arise by pushing applications that are dependent on internal network-level security controls to the cloud.

Organizations might also face unknown operational and architectural issues when they use designers and architects that are unfamiliar with cloud technologies for their applications. Operational issues like accounts management and user rights management can hinder the proper operation of the organization's system on the cloud [21].

A worse case is when an unprepared organization take the decision to create their own private cloud. While some organizations make a conscious decision to create their own cloud without considering the requirements, other organization are forced by law to keep their data in the house. Government organizations, for example, in many countries are forced to keep their sensitive and private data inside their premises.

It feels much safer to keep your data in your organization. However, this feeling might be false security if your organization is not technically capable of handling a privately owned cloud.

2.9 Shared Technology Vulnerabilities

At the core of cloud computing, many technologies collaborate to provide the cloud services. Any vulnerability existing at the backend can lead to full-system exploitation in all clients. There were cases where the underlying architecture (such as CPU caches, GPUs, etc.) does not provide complete isolation properties. This can enable the attacker to use one client virtual machine to exploit other clients' virtual machines.

When an integral part is compromised, such as the hypervisor, it can expose the entire cloud environment.

Although this threat is considered dangerous because it can affect a complete cloud all at once, its severity have dropped over the past few years. This drop is due to more accurate configuration and isolation by the hardware manufacturers and the cloud service providers. It is expected that the future developments in hardware will drive toward more isolation between hosted virtual machines. This increased isolation will be very beneficial in terms of security.

2.10   Other Threats

This list includes other threats to cloud computing that are thought of as less dangerous and were not included in the Notorious Nine report in [28]:

1. Lock-in: The inability to change the cloud service provider [11].
2. Insecure or Incomplete Data Deletion [11].
3. Threats to Trust: in the context of cloud computing it focuses on convincing observers that a system (model, design, or implementation) was correct and secure [20].
4. Loss of Governance: There are some cases in which the service agreements do not cover all security aspects such that the client is unclear about which measures need to be taken by the client and which by the cloud service provider [11].
5. Difficulty in Forensics Analysis After Security Breaches [14].
6. Acquisition of the cloud provider by another organization might induce a strategic shift in the service provider strategy and may put nonbinding agreements at risk of cancellation or change [11].

More details about other threats can be found in [4, 5, 18, 19].

3   How Is Cloud Security Different

Cloud computing systems include their share of vulnerabilities just like any other type of systems. These vulnerabilities, when exploited by attackers, can cause service disruptions, data loss, data theft, ... etc. Given the unique nature of dynamic resource allocation in the cloud, it is possible that classical attacks and vulnerabilities can cause more harm on a cloud system if it is not protected properly.

Nothing explains how the cloud is different better than an example. One of the unique characteristics of the cloud is availability. The cloud is designed to be available all the time. Whether it is a private or a public cloud, availability is an undeniable feature that many organizations seek. What if attackers target availability of the cloud?

One of the major reasons why organizations decide to switch to a cloud environment is the "you-pay-for-what-you-use" business model. No one likes paying for resources that are not very well utilized. Hence, when an attack such as Denial-of-Service (DoS) happens, not only availability is targeted.

DoS attacks aim at making a certain network service unavailable to its legitimate users. In its basic form, this attacks keep the resources busy such that these resources become unavailable to the users this service was aimed to serve.

Using DoS attacks on the cloud, the attacker can cause huge financial implications by consuming high resources in the trial of making the service unavailable. So, for the organization using the cloud, its a doubled loss.

The organization will be paying a lot of money for the resources consumed by the attack and, after a while, the organizations service will be unavailable due to the

DoS attack. This type of attacks is referred to as Fraudulent Resource Consumption (FRC)[17].

This is why the uniqueness of the cloud technology open the door for unique attacks or at least unique effects of old common attacks. Keep in mind that the huge processing power of the cloud can be used by attackers as a powerful attacking tool. The attacker can benefit from the virtually unlimited on-demand processing power and employ it in performing DoS attacks among other attacking choices. In the coming subsections, we will discuss the main difference points between a cloud-based system and a classical data-center model.

3.1   New Attack Surfaces

Looking at the cloud architecture shown in Fig. 1, we can easily see that any new layer within the architecture means a new attacking surface. The abstraction layer, for example, can be a target for a whole new class of attacks.

Having the multiple layers shown earlier in Fig. 1, cloud computing can be target for attacks at any of these levels. Threats exist at virtually any level of a cloud computing system. As you have seen in the previous section, there are threats at the hypervisor level, threats at the platform level, threats at the software level, ...etc. Many of the attacks that try to exploit these threats are unique to the cloud and cannot be used on classical data-center security model.

The abstraction layer software, often called the hypervisor, can be a target for malicious attacks. As shown in Fig. 1, the hypervisor sits between the hardware and the VMs that comprise the cloud. Although not many attacks were conducted on hypervisors before, any compromise in the hypervisor security can bring the whole cloud down [5].

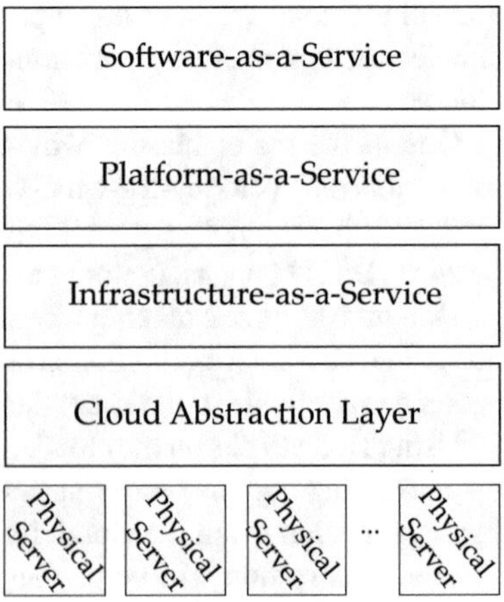

Fig. 1   Cloud computing architecture

Hyperjacking was identified in [23] as the attackers attempt to craft and run a very thin hypervisor that takes complete control of the underlying operating system. Once the attacker gains full control of the operating system, the whole cloud is compromised. The attacker will be able to eavesdrop, manipulate clients data, disrupt, or even shut down the complete cloud service. Although the probability of this attack succeeding is very low, it is still a source of concern.

A hypervisor vulnerability was previously reported in [8]. This vulnerability was found in many commercial VM and cloud computing products. The vulnerability enables privilege escalation from guest account privileges to host account privileges.

3.2 Different Architecture Means Different Security Measures

Clearly, cloud architecture is different. Looking at Fig. 2 we can see the differences in architecture between a classical server, virtualized server, and cloud-based server systems.

The difference in architecture implies difference in security measures. For example, in a data-center security model, it is clear where you need to place the Intrusion Detection System (IDS) or a Firewall. However, in a cloud-based system, there is no clear line of defense where you can place a firewall. Should it be placed on the border of the whole cloud network? should it be placed at the border of every physical machine? should it be place at the border of every virtual machine?

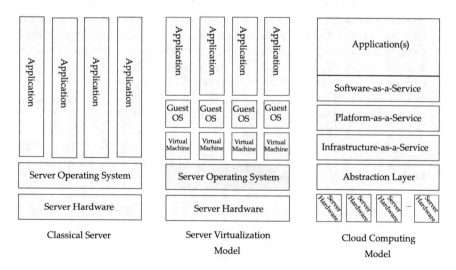

Fig. 2 Comparison of classical server, virtualized server, and cloud-based server systems

What happens when one client in the cloud requires more security than others? should the VMs of this client be on an isolated network? wouldn't that limit the cloud's flexibility, which is one of the most important features of the cloud?

Many questions need to be addressed properly so that the limits of responsibility can be defined.

3.3   Limits of Responsibility

The limits of responsibility distribution between the client organization and the cloud service provider depends mainly on the cloud service model. The different service models identify the level of involvement of the client in the server provisioning process. The higher you go in Fig. 1 the less involved the client needs to be and the higher the cost becomes. Whenever you give more work to the service provider, the cost meter tends to go up.

Generally, the three service models are:

1. Infrastructure-as-a-Service
   IaaS service model is the lowest level of service provided to the client. In this service model, the cloud computing client is provided with controlled access to the virtual infrastructure. Using this access, the client can install operating system and application software. From the client's point of view, this model is similar to renting the hardware from a service provider and letting the service provider manage the hardware. In this sense, the client does not have control over the physical hardware. On the other hand, the client will have to manage the security aspects from the operating system and up to the applications. This model requires the client to have highly experienced network engineer(s). Handling everything from the operating system and up is a big responsibility that most clients decline to handle, especially because of the security burdens. Thus, this model is not of high preference in the cloud computing clients' society [15].
   In summary, IaaS takes away the burden of procurement and maintenance of hardware and pushes it over to the cloud service provider side.
2. Platform-as-a-Service
   In PaaS, the operating system and all platform-related tools (like compilers) are already installed for the client. These pre-installed components are also managed by the cloud service provider.
   Clients have the freedom of installing additional tools based on their needs. However, the control over the infrastructure is retained by the service provider. The client controls applications development, configuration, and deployment. In some aspects, this service model is similar to the traditional web-hosting services in which clients rent a remote server with development platform pre-installed on it. The major difference between this model and traditional web-hosting is the rapid provisioning. Traditional web-hosting is managed manually and requires human intervention when the demand increases or decreases. On the other hand,

provisioning in cloud computing is automatic and rapid. Thus, it does not require any human interventions [15].

3. Software-as-a-Service

   SaaS model focuses on the application level and abstracts the user away from infrastructure and platform details. Usually, applications are provisioned via thin client interfaces such as web browsers or even mobile phone apps [15]. Microsofts Outlook.com is a clear example of this. An organization can adopt Outlook.com electronic mail service and never bother with hardware mainte-nance, service uptime, security, or even operating system management. The client is given the control over certain parameters in the software configuration, for example, creating and deleting mail boxes. These parameters can be controlled through the interface of the application.

   This service model gives the client the luxury of not worrying about hardware, operating system, host security, patching and updating,...etc. Instead, the client will be able to focus on using the application and achieving business goals.

Although service model is a deciding factor of the limit of responsibilities of each party, there is another factor that plays an important role; Service-Level Agreement (SLA). Many details about the security of the cloud system and liabilities of the service provider are usually written in the SLA.

4   Prioritizing Assets and Threats

As the case with any system, managing the security processes is as important as the security processes themselves. The first step in any security project is risk assessment [13]. Preparations for risk assessment includes identifying the organization's assets.

Before starting the risk assessment, it is a good idea to focus on assets. We can categorize these assets into 5 types:

1. Hardware assets
2. Software assets
3. Data
4. Human resources
5. Business goals

Yes, I would like to consider business goals as assets as well because I can have all the other assets and fail as a business if I did not achieve the business goals and did not maintain business continuity.

Assets need to be valuated properly so that you can make proper decisions regard-ing the assets' security. This valuation process is vital and we will explain its impor-tance with an example. Lets say that you have 10 different types of cars. You have a budget to rent 5 covered parking spots and 5 other spots under the sun. How would you decide which car to put in the sun and which car goes into the shaded area?

There are a few deciding factors; the most important one is the cost of the car. You would want to put your most valuable 5 cars in the shade and let the other 5 park in the sunny area.

After creating a list of all of your assets and distributing them into different categories you will need to look into your business goals and see which assets are more vital in achieving your business goals. You might end up leaving a Porche in the sun to protect a pick-up truck that you use a lot in business.

Once you have categorized and valuated your assets, you can start with the risk assessment process. The risk assessment is done in the following steps [25]:

1. Identify threat sources and events
2. Identify vulnerabilities and predisposing conditions
3. Determine likelihood of occurrence
4. Determine magnitude of impact
5. Determine risk

These steps would help the cloud computing client organization in identifying suitable cloud solutions. Some cloud service providers can provide techniques to reduce the probability of these risks. By prioritizing these risks, the organization will have the ability to decide how much it is willing to spend to protect each asset.

In addition to deciding the spending budget to secure the assets, the organization will also be able to decide which cloud service provider to choose and entrust for their systems and data.

Part of that decision is based on identifying the cost of the security controls needed to mitigate each risk. Security controls can be divided into [13]:

1. Directive: Controls designed to specify acceptable rules for behavior within an organization.
2. Deterrent: Controls designed to discourage people from violating security directives.
3. Preventive: Controls implemented to prevent a security incident or policy violation.
4. Compensating: Controls implemented to substitute for the loss of primary controls and mitigate risk down to an acceptable level.
5. Detective: Controls design to signal a warning when a security control has been breached.
6. Corrective: Controls implemented to remedy circumstances, mitigate damage, or restore controls.
7. Recovery: Controls implemented to restore conditions to normal after a security incident.

The cost of each control varies depending on the type of the control and the asset that it is designed to protect. After determining the cots of the required controls, you will have the ability to decide which ones to adopt to mitigate or eliminate risks and which risks to accept.

To put things into perspective, we will consider a calculation model used to support decisions on which security control is worth paying for and which is less important. The reason why we are focusing more on the managerial aspect rather than the technical aspect is that in almost all cases the budget is not adequate to provide all security controls required to cover all threats and eliminate all risks.

Generally, we can say that

$$r = (p \cdot v) - m + u \tag{1}$$

where,

p is the probability of the threat succeeding to achieve an undesirable effect.

v is the value of the asset.

m is the percentage of risk mitigated by the currently existing controls (0% if no controls exist).

u is the percentage of uncertainty in the information available about the vulnerability and threat.

To understand this better, we will look into an example. Lets say that we have an asset named A. This asset's value is 130. This asset has a vulnerability named Vulnerability#1. This vulnerability has a probability of success of 0.45. Currently, there is no control mitigating this risk. The trust we have in this information is only 70%. This means that the uncertainty is 100% − 70% = 30%. Now, lets calculate the risk:

$$r = (0.45 \times 130) - 0\% + 30\% \tag{2}$$

$$r = 58.5 + 30\% = 76.05 \tag{3}$$

This number alone is not very useful without putting it into a table and looking into all the numbers from all the vulnerabilities in all the assets. Hence, we can develop a table to put all of this information and re-arrange the table to put the highest risk on the top and the lowest risk below. Then, we add more columns to that table which is the new calculations after adding the suggested security controls to see the effect of adding these controls in reducing the risks. This would help the management in understanding the budget that needs to be allocated to information security. The final table is expected to look like the sample table shown in Table 1.

Looking at Table 1, we can see that some suggested security controls will be very effective in reducing the risk like Asset A Vulnerability#1 and Asset B Vulnerability#1. Some other suggested controls would have a moderate effect like Asset A Vulnerability#2 and Asset B Vulnerability#2. While in the case of Asset A Vulnerability#3, the suggested security control had a limited effect in reducing the risk.

This type of analysis will help us in prioritizing assets and prioritize spending on security measures to protect out data on the cloud. By combining the knowledge of threats available in cloud computing with the calculations done in this section, we can produce the best possible security plan with a given budget.

Table 1 Sample table of risk calculations

Asset	Asset value	Vuln#	Success probability	Current			Suggested			Reduction
	v		p	m	u	r	m	u	r	
A	145	1	0.30	0%	20%	52.2	60%	10%	21.75	30.45
		2	0.15	50%	10%	13.05	90%	5%	3.2625	9.7875
		3	0.10	0%	5%	15.225	30%	5%	10.875	4.35
B	80	1	0.4	0%	10%	35.2	75%	5%	9.6	25.6
		2	0.25	30%	20%	18	60%	10%	10	8

5 Cloud Security Considerations

The following list provides general considerations for cloud computing security:

- Read the SLA carefully.
  Know the responsibilities of the service provider to have a clear understanding of what you will need to do. SLAs are the documents to go back to whenever anything goes bad. When you are selecting a cloud service provider, choose the one that suites your technical capabilities. Do not consider only the cost as a deciding factor.
- Know the cloud service provider's history.
  A cloud service provider that has been a target for repeating attacks is not a good choice. It might not be possible to find a cloud service provider that has not been targeted before, but it is always a good idea to choose the one that responds well in crisis.
- Do not use vendor supplied defaults for passwords and other security parameters [5].
  The default usernames, passwords, and other security parameters are well known to other cloud users, and probably many outsiders. It is vital that the first step done when you receive your account credentials that you change passwords for all accounts.
- Do not setup your own private cloud unless it is absolutely necessary [5].
  Scenarios where the organization decides to build it's own private cloud has to be limited by all of the following conditions:

1. The organization is obliged by some governance laws or guidelines that force them to host their user data inside their premises.
2. The organization has adequate budget for the project. As we have explained in Chap. 1, the capital expenditures for starting a private cloud are very high.
3. In the case that the organization, for security purposes, intends to build their private cloud, they will have to provide the human resources needed for the job. In some looser scenarios, it would be possible for the organization to outsource the creation of the cloud and be responsible for cloud maintenance only. In that case, the organization must have the human resources capable of maintaining the cloud.
4. The organization must be capable of creating and maintaining a proper incident response plan, disaster recovery plan, and business continuity plan, along with all their requirements (e.g.,: backup generators, backup Internet connection, alternate location,...etc)

  Unless all of the aforementioned conditions exist, it is recommended that the organization considers public cloud service providers or alternative solutions like community cloud, or even a hybrid cloud solution.
- Security patches reduce the risks of zero-day attacks.
  A zero day attack is an attack which exploit vulnerabilities that have not been

disclosed publicly [7]. Methods of defense against these attacks are not available until they are detected publicly. Among the findings that [7] presented were:

1. Zero-day attacks are more frequent than previously thought: 11 out of 18 vulnerabilities identified were not known zero-day vulnerabilities.
2. Zero-day attacks last between 19 days and 30 months, with a median of 8 months and an average of approximately 10 months.
3. After zero-day vulnerabilities are disclosed, the number of malware variants exploiting them increases 18,385,000 times and the number of attacks increases 2,100,000 times.
4. Exploits for 42% of all vulnerabilities employed in host-based threats are detected in field data within 30 days after the disclosure date.

These findings tell us how important it is to keep all our systems patched and updated. Keep in mind that these findings are for the years 2008–2011. However, they give us a general idea of how bad zero-day attacks are. As there is not much to do in defense from zero-day attacks, patching in a timely manner is important. The security personnel in the organization need to be up-to-date on all new vulnerabilities, threats, and attacks.

References

1. Nasdaq Server Breach: 3 Expected Findings. Retrieved March 29, 2016, from http://www.darkreading.com/attacks-and-breaches/nasdaq-server-breach-3-expected-findings/d/d-id/1100934?.
2. Zeus Bot Found Using Amazons EC2 as C and C Server. Retrieved March 30, 2016, from http://goo.gl/g9PCtQ, a.
3. Amazon Purges Account Hijacking Threat from Site. Retrieved March 30, 2016, from http://goo.gl/JJqxtd, b.
4. Alani, M. M. (2014). Securing the cloud: Threats, attacks and mitigation techniques. *Journal of Advanced Computer Science & Technology, 3*(2), 202.
5. Alani, M. M. (2016). *Elements of cloud computing security: A survey of key practicalities.* London, UK: Springer.
6. Almorsy, M., Grundy, J., & Ibrahim, A. S. (2011, July). Collaboration-based cloud computing security management framework. In *2011 IEEE International Conference on Cloud Computing (CLOUD)* (pp. 364–371).
7. Bilge, L., & Dumitras, T. (2012). Before we knew it: An empirical study of zero-day attacks in the real world. In *Proceedings of the 2012 ACM Conference on Computer and Communications Security* (pp. 833–844). ACM.
8. CERT Vulnerability Note VU. Vulnerability note vu no 649219.
9. Chong, F., Carraro, G., & Wolter, R. (2006). *Multi-tenant data architecture* (pp. 14–30). Microsoft Corporation, MSDN Library.
10. Chow, R., Golle, P., Jakobsson, M., Shi, E., Staddon, J., Masuoka, R., et al. (2009). Controlling data in the cloud: Outsourcing computation without outsourcing control. In *Proceedings of the 2009 ACM Workshop on Cloud Computing Security* (pp. 85–90). ACM.
11. European Network and Information Security Agency. *Cloud computing: Benefits, risks and recommendations for information security.* ENISA (2009).

12. Galante, J., Alpeyev, P., & Yasu, M. (2016). Amazon.com server said to have been used in sony attack. Retrieved September 28, 2016, from, http://www.bloomberg.com/news/articles/2011-05-13/sony-network-said-to-have-been-invaded-by-hackers-using-amazon-com-server.
13. Gordon, A. (2015). *Official (ISC) 2 Guide to the CISSP CBK*. CRC Press.
14. Grispos, G., Storer, T., & Glisson, W. B. (2013). Calm before the storm: The challenges of cloud. *Emerging Digital Forensics Applications for Crime Detection, Prevention, and Security*, *4*(1), 28–48.
15. Hill, R., Hirsch, L., Lake, P., & Moshiri, S. (2012). *Guide to cloud computing: Principles and practice*. Springer Science & Business Media.
16. Hong, J. (2012). The state of attacks. *Communications of the ACM*, *55*(1), 74–81.
17. Idziorek, J., Tannian, M., & Jacobson, D. (2012). Attribution of fraudulent resource consumption in the cloud. In *2012 IEEE 5th International Conference on Cloud Computing (CLOUD)* (pp. 99–106). IEEE.
18. Kazim, M., & Zhu, S. Y. (2015). A survey on top security threats in cloud computing. *International Journal of Advanced Computer Science and Applications (IJACSA)*, *6*(3), 109–113.
19. Khalil, I. M., Khreishah, A., & Azeem, M. (2014). Cloud computing security: A survey. *Computers*, *3*(1), 1–35.
20. Nagarajan, A., & Varadharajan, V. (2011). Dynamic trust enhanced security model for trusted platform based services. *Future Generation Computer Systems*, *27*(5), 564–573.
21. Nair, S. K., Porwal, S., Dimitrakos, T., Ferrer, A. J., Tordsson, J., Sharif, T., et al. (2010). Towards secure cloud bursting, brokerage and aggregation. In *2010 IEEE 8th European Conference on Web Services (ECOWS)*, December 2010 (pp. 189–196). doi:10.1109/ECOWS. 2010.33.
22. Ramgovind, S., Eloff, M. M., & Smith, E. (2010). The management of security in cloud computing. In *2010 Information Security for South Africa*, August 2010 (pp. 1–7).
23. Ray, E., & Schultz, E. (2009). Virtualization security. In *Proceedings of the 5th Annual Workshop on Cyber Security and Information Intelligence Research: Cyber Security and Information Intelligence Challenges and Strategies* (p. 42). ACM.
24. Rittinghouse, J. W., & Ransome, J. F. (2016). *Cloud computing: Implementation, Management, and Security*. CRC Press.
25. Ross, R. S. (2011). *Guide for conducting risk assessments* (pp. 800–830). NIST Special Publication.
26. Shirey, R. (2000). RFC 2828: Internet security glossary. *The Internet Society*, *13*.
27. Takabi, H., Joshi, J. B. D., & Ahn, G.-J. (2010). Security and privacy challenges in cloud computing environments. *IEEE Security & Privacy*, *6*, 24–31.
28. Top Threats Working Group et al. (2013). The notorious nine: Cloud computing top threats in 2013. *Cloud Security Alliance*.
29. Veriato. (2016). Insider threat spotlight report. Retrieved March 30, 2016, from http://goo.gl/rcGKcQ.
30. Zhang, Y., Juels, A., Oprea, A., & Reiter, M. K. (2011). Homealone: Co-residency detection in the cloud via side-channel analysis. In *2011 IEEE Symposium on Security and Privacy (SP)* (pp. 313–328). IEEE.
31. Zhang, Y., Juels, A., Reiter, M. K., & Ristenpart, T. (2012). Cross-VM side channels and their use to extract private keys. In *Proceedings of the 2012 ACM Conference on Computer and Communications Security* (pp. 305–316). ACM.
32. Zhou, Y. B., & Feng, D. G. (2005). Side-channel attacks: Ten years after its publication and the impacts on cryptographic module security testing. *IACR Cryptology ePrint Archive*, *2005*, 388.

An Overview of Cloud Forensics Strategy: Capabilities, Challenges, and Opportunities

Reza Montasari

Abstract Cloud computing has become one of the most game changing technologies in the recent history of computing. It is gaining acceptance and growing in popularity. However, due to its infancy, it encounters challenges in strategy, capabilities, as well as technical, organizational, and legal dimensions. Cloud service providers and customers do not yet have any proper strategy or process that paves the way for a set procedure on how to investigate or go about the issues within the cloud. Due to this gap, they are not able to ensure the robustness and suitability of cloud services in relation to supporting investigations of criminal activity. Moreover, both cloud service providers and customers have not yet established adequate forensic capabilities that could assist investigations of criminal activities in the cloud. The aim of this chapter is to provide an overview of the emerging field of cloud forensics and highlight its capabilities, strategy, investigation, challenges, and opportunities. This paper also provides a detailed discussion in relation to strategic planning for cloud forensics.

1 Introduction

Cloud computing has become one of the most transformative computing technologies, following the footsteps of main-frames, minicomputers, personal computers, the World Wide Web, and smartphones [25, 29]. Cloud computing is drastically transforming the way in which information technology services are created, delivered, accessed, and managed. Spending on cloud services is growing at five times the rate of traditional on-premises information technology (IT). Cloud computing services are forecast to generate approximately one-third of the net new growth within the IT industry. Just as the cloud services market is growing, at the same time the size of the average digital forensic case is growing. This culminates

R. Montasari (✉)
Department of Computing and Mathematics, University of Derby,
Kedleston Road, Derby DE22 1GB, UK
e-mail: r.montasari@derby.ac.uk

© Springer International Publishing AG 2017
A. Hosseinian-Far et al. (eds.), *Strategic Engineering for Cloud Computing and Big Data Analytics*, DOI 10.1007/978-3-319-52491-7_11

189

in the amount of forensic data that must be processed, outgrowing the ability to process it in a timely manner [27]. The growth of cloud computing not only aggravates the problem of scale for digital forensic practice, but also generates a brand new front for cybercrime investigations with the related challenges [29]. Cloud forensic investigators, hereafter referred to as CFIs, should spread their expertise and tools to cloud computing environments. In addition, cloud service providers, hereafter referred to as CSPs, and cloud customers must create forensic capabilities that can assist with reducing cloud security risks. To this end, this chapter discusses the emerging area of cloud forensics, and highlights its capabilities, strategy, investigations, challenges, and opportunities.

The chapter is structured as follows: Sect. 2 discusses the three different dimensions of cloud forensics. Section 3 describes the way in which forensic investigations are carried out in the cloud and discusses the practice of cloud forensics. In Sect. 4, the major challenges posed to the cloud forensic investigations are discussed, and in Sect. 5, some of the opportunities presented by cloud forensics are presented. Finally, the chapter is concluded in Sect. 6.

2 The Three Aspects of Cloud Forensics

Cloud forensics is a cross discipline of cloud computing and digital forensics [29]. Cloud computing is a shared collection of configurable networked resources (e.g., networks, servers, storage, applications, and services) that can be reconfigured quickly with minimal effort [19], while digital forensics is the application of computer science principles to recover electronic evidence for presentation in a court of law [15, 21]. Cloud forensics is also considered to be a subdivision of network forensics. Network forensics deals with forensic investigations of networks, while cloud computing deals with broad network access. Hence, cloud forensics is based on the main phases of network forensics with techniques tailored to cloud computing environments. Cloud computing is a developing phenomenon with complex aspects. Its vital features have considerably reduced IT costs, contributing to the swift adoption of cloud computing by business and government [11]. To ensure service availability and cost effectiveness, cloud service providers maintain data centers around the world. Data stored in one data center is replicated at multiple locations to ensure abundance and reduce the risk of failure [29]. Also, the separation of responsibilities between CSPs and customers in relation to forensic responsibilities vary in accordance with the service models being employed. Analogously, the interactions between different tenants that employ the same cloud resources vary in accordance with the deployment model being used. Multiple jurisdictions and multi-tenancy are the default settings for cloud forensics creating further legal issues. Sophisticated relations between CSPs and customers, resource sharing by multiple tenants and collaboration between international law enforcement agencies are required in most cloud forensic investigations. In order to examine the domain of cloud forensics in more detail and to highlight that cloud forensics is a multi-dimensional issue rather than simply a technical issue, this paper discusses technical, organizational, and legal aspects of cloud forensics next.

2.1 Technical Aspects

The technical dimension involves a set of tools and procedures required to conduct the forensic process in cloud computing environments [10, 29]. These comprise of proactive measures (forensic readiness) [21, 28, 31], data collection, elastic/static/live forensics [12, 21], evidence segregation, and investigations in virtualized environments [10, 29]. Data acquisition is the process of identifying, acquiring, and verifying digital data that might represent potential digital evidence [14]. In the context of cloud forensics, digital data encompasses client-side artifacts stored on client premises, as well as provider-side artifacts residing in the provider infrastructure. The procedures and tools employed to acquire forensic data vary according to the specific model of data responsibility that is in place [18, 33]. During the acquisition process, the integrity of data should be preserved according to clearly defined division of responsibilities between the client and provider. While conducting the data acquisition, CFIs will need to ensure that they do not infringe on laws or regulations in the jurisdiction in which digital data is being acquired. Moreover, they also need to ensure that they do not compromise the confidentiality of other tenants who use the same resources. For instance, in public clouds, provider-side artifacts might need the separation of tenants, while there might be no such requirement in private clouds [29].

There are five essential characteristics of cloud computing including: (1) on-demand self-service, (2) broad network access, (3) measured service, (4) rapid elasticity, and (5) resource pooling [9, 19]. In relation to the 'on-demand self-service', a consumer can unilaterally provision computing capabilities, such as server time and network storage, as required automatically without requiring human interaction with each service provider [19]. Concerning the 'broad network access', capabilities are available over the network and can be accessed through standard mechanisms that promote use by "heterogeneous thin or thick client platforms" (e.g., mobile phones, tablets, laptops, and workstations) [19]. With regards to 'measured service', cloud systems automatically control and improve resource use by taking advantage of a metering capability at some level of abstraction appropriate to the type of service (e.g., storage, processing, bandwidth, and active user accounts). Resource usage can be monitored, controlled, and reported; this provides clarity for both the provider and consumer of the deployed service.

In terms of rapid elasticity, capabilities can be elastically provisioned and released, in some cases automatically, to scale rapidly outward and inward in proportion with demand. Therefore, it is necessary that cloud forensic tools are also elastic [9, 29]. These often consist of large-scale static and live forensic tools for data collection, data recovery, evidence examination, evidence analysis, evidence interpretation, event reconstruction, and evidence presentation [7, 16, 21]. Another vital trait of cloud computing is resource pooling, in which the provider's computing resources are pooled to serve multiple consumers utilizing a multi-tenant model, with different physical and virtual resources dynamically allocated and reallocated based on consumer demand [9, 19]. Multi-tenant environments reduce IT costs through resource sharing. However, the process of separating evidence in the cloud necessitates compartmentalization [5]. Thus, procedures and tools must be created to separate forensic data between multiple tenants in various cloud deployment models and service models [29]. Therefore, there exists a sense of location independence in that the customer generally does not have control or knowledge over the exact location of the provided resources. However, they might be able to specify location at a higher level of abstraction (e.g., country, state or datacenter). Examples of resources include storage, processing, memory, and network bandwidth [9].

2.2 Organizational Aspects

At least two parties are always involved in forensic investigations in cloud computing environments. These include CSP and cloud customer [10, 29]. However, when the CSP outsources services to other parties, the scope of the investigation tends to widen. CSPs and most cloud applications often have dependencies on other CSPs. The dependencies in a chain of CSPs/customers can be highly dynamic. In such circumstances, the CFI may depend on investigations of each link in the chain [29]. If there is any interruption or corruption in the chain or a lack of coordination of duties between all the involved parties, this can culminate in serious problems. Organizational policies and service level agreements (SLAs) pave the way for communication and collaboration in cloud forensic practice. Moreover, law enforcement and the chain of CSPs must communicate and co-operate with third parties and academia. Third parties can aid auditing and compliance while academia can provide technical expertise that could improve the efficiency and effectiveness of investigations [10]. Therefore, when establishing a cloud forensic strategy of an organization to investigate cloud anomalies, each CPS needs to create a permanent or ad hoc department that would be in charge of internal and external matters that must accomplish the following roles:

Cloud Forensic Investigators
CFIs must examine claims of misconduct and work with external law enforcement agencies as required. They will need to have adequate expertise in order to carry out investigations of their own assets as well as interact with other parties in CFIs.

IT Professionals
IT professionals are comprised of system, network and security administrators, ethical hackers, cloud security architects, and technical and support staff [29]. They must provide expert knowledge to support investigations, help investigators access digital crime scenes, and might carry out data acquisition on behalf of investigators.

First Responders

First responders handle security incidents such as unauthorized data access, accidental data leakage and loss, breach of tenant confidentiality, inappropriate system use, malicious code infections, insider attacks, and denial of service attacks [13, 29]. All cloud entities should have strategy (written plans or Standard Operating Procedures) that classify security incidents for the different levels of the cloud and identify first responders with the relevant expertise.

Legal Practitioners

Legal practitioners are accustomed to multi-jurisdictional and multi-tenancy issues in the cloud. They have the responsibility for ensuring that cloud forensic practice does not infringe on laws and regulations. They also have a duty to uphold the confidentiality of other tenants that share the resources. SLAs must elucidate the procedures followed in cloud forensic investigations. Internal legal practitioners must be part of the team that drafts the SLAs in order to cover all the jurisdictions where a CSP operates. Internal legal practitioners also have a duty to communicate and co-operate with external law enforcement agencies during the course of cloud forensic investigations [29].

External Assistance

It is also judicious for a CSP to depend upon internal staff in addition to external parties to carry out forensic activities. It is vital for a CSP to define the actions in advance.

2.3 Legal Aspects

Multi-jurisdictional and multi-tenancy challenges must be considered as the top legal concerns [4, 17, 22]. Conducting forensic investigations in the cloud aggravates these challenges [10, 29]. The legal aspect has various dimensions. The first dimension is the multi-jurisdiction and multi-tenancy challenges considered as top-level concerns among digital forensic experts, and are both exacerbated by the cloud. Regulations and agreements must be maintained in the legal aspect of cloud forensics so as to ensure that the investigations will not infringe on any laws or regulations in the area in which the data physically resides [18]. Moreover, steps must also be taken to ensure that the privacy of other individuals or organization sharing the infrastructure will not be jeopardized or compromised throughout the forensic practice. Another dimension of the legal aspect is the SLA that defines the terms of use between the cloud customer and the CSP. Ruan et al. [29] provide a list of terms that needs to be included in the existing SLAs so as to assist with making cloud forensic investigations. These include

- The customers must be granted services, access, and techniques by the CSP during digital forensic investigation.
- Trust boundaries, roles, and responsibilities between the customer and the CSP must be clearly defined during forensic investigation.

- Legal regulations and relevant laws must be addressed during a multi-jurisdictional forensic investigation, as well as in a multi-tenant environment.
- Confidentiality of customer data and privacy policies must be taken into account during a multi-jurisdictional forensic investigation, as well as in a multi-tenant environment.

3 Practice of Cloud Forensics

Cloud forensics has several main uses, including: forensic investigation, troubleshooting, log monitoring, data and system recovery, and due Diligence/Regulatory Compliance. The followings are the main five usages of the cloud forensics as outlined by Cruz [10]:

Investigation

- On cloud crime and policy violations in multi-tenant and multi-jurisdictional environments
- On suspect transactions, operations, and systems in the cloud for incident response
- Event reconstructions in the cloud
- On the acquisition and provision of admissible evidence to the court
- On collaborating with law enforcement in resource confiscation.

Troubleshooting

- Finding data and hosts physically and virtually in cloud environments
- Determining the root cause for both trends and isolated incidents, as well as developing new strategies that will help prevent similar events from happening in the future
- Tracing and monitoring an event, as well as assessing the current state of said event
- Resolving functional and operational issues in cloud systems
- Handling security incidents in the cloud.

Log Monitoring

- Collection, analysis, and correlation of log entries across multiple systems hosted in the cloud, including but not limited to: audit assists, due diligence, and regulatory compliance.

Data and System Recovery

- Recovery of data in the cloud, whether it has been accidentally or intentionally modified or deleted
- Decrypting encrypted data in the cloud if the encryption key is already lost

- Recovery and repair of systems damaged accidentally or intentionally
- Acquisition of data from cloud systems that are being redeployed, retired or in need of sanitation.

Due Diligence/Regulatory Compliance

- Assist organizations in exercising due diligence as well as in complying with requirements related to the protection of sensitive information, maintenance of certain records needed for audit, and notification of parties concerned when confidential information is exposed or compromised.

Notice that describing each of the above five usages of the cloud forensics is outside the scope of this chapter. Therefore, due to the space constraint, only the investigation aspect is discussed in detail within the following section.

3.1 Crime and Cloud Forensic Investigations

Cloud crime can be defined as any crime that includes cloud computing in the sense that cloud can be the subject, object, or tool related to the crimes [10]. The cloud is considered to be the object when the target of the crime is the CSP that is directly targeted by the criminal act, such as by Distributed Denial of Service (DDoS) attacks that affect sections of the cloud or the cloud itself as a whole. The cloud can also be considered as the subject of the criminal activity when the criminal act is perpetrated within the cloud environment, for instance in cases of identity theft of cloud users' account. Moreover, cloud can also be considered the tool when it is employed to plan or carry out a crime such as in cases when evidence associated with the crime is stored and distributed in the cloud, or a cloud is employed to attack other clouds. Cloud forensics is the intersection between cloud computing and network forensic analysis. As already stated, cloud computing refers to a network service that users can interact with over the network. This denotes that all the work is carried out by a server on the Internet, which might be supported by physical or virtual hardware. In recent years, there has been a significant growth on the deployment of virtualized environments, which makes it very likely that the cloud service is running somewhere in a virtualized environment [18].

Although cloud computing has many benefits, at the same time it also poses two most distinct disadvantages including 'security' and 'privacy'. Since the users' data are stored in their cloud on the Internet, the CPS has access to that data, and so does an attacker (if a breach occurs in the provider's network). If a breach has occurred and it has been decided to collect data, the user must decide why the data are being acquired (e.g., for remediation, court, or some other reason) and, thus, what data need to be collected [32]. If the user is able to acquire the necessary data in the normal course of business through the company's access to the cloud, they must then revert to the standard digital forensic techniques by following well-established methods such as those presented in Montasari [21], Montasari et al. [22], ACPO [1].

If the case is to be presented in court of law, they will also need to ensure that they will maintain a strict and an accurate chain of custody through a detailed documentation [7, 21, 31]. However, standard digital forensic techniques might not be sufficient in certain cases, thus necessitating network forensic analysis techniques. Network forensic analysis is a branch of digital forensics, which screens and analyzes computer network traffic for the purposes of gathering information, collecting legal evidence, or detecting intrusions [21, 22, 24]. Network forensics refers to the data transmitted over the network, which might serve as the only evidence of an intrusion or malicious activity. Obviously this is not always the case due to the fact that an intruder often leaves behind evidence on the hard disk of the compromised host, in log files and also uploaded malicious files, etc. [2].

In circumstances where the intruder has been able to avoid leaving any artifact on the compromised computer, the only evidence that might be available is in the form of captured network traffic [6]. When capturing network traffic, DFI will need to separate the good data from the bad by extracting useful information from the traffic, such as transmitted files, communication messages, credentials, etc. [16]. If DFIs are faced with a large volume of disk space, they must store all the traffic to disk and analyze it at a later time if needed. However, this obviously requires a vast amount of disk space [26]. Usually network forensics is deployed to discover security attacks being conducted over the network. A tool such as Tcpdump or Wireshark can be used to carry out the network analysis on the network traffic [18]. Cloud network forensics is necessary in circumstances in which attackers have hacked the cloud services. In these situations, the CFIs must look through all the logs on the compromised service in an attempt to recover forensic artifacts. It might be the case that the CFIs discover that the attack was carried out from the cloud provider's network. In such circumstance, they must request the cloud provider to give them the logs that they require. At this juncture, it is important to evaluate what logs the CFIs require in order to identify who carried out the attack. This is where cloud network forensics plays a vital role.

CFIs will need to apply a Digital Forensic Investigation Process Model (DFIPM) such as that proposed in Montasari [21] to the cloud, where the CFIs would need to analyze the information they have concerning filesystems, processes, registry, network traffic, etc. When gathering information that they can examine and analyze, CFIs must be aware of which service model is in usage due to the fact that acquiring the appropriate information depends upon it.

When using different service models, CFIs can access different types of information as shown in the Table 1 [18, 33]. The first column consists of different layers that CFIs can access when deploying cloud services. The SaaS, PaaS, and IaaS columns represent the access rights that the CFIs are granted when employing various service models. The last column shows the information that the CFIs have available when employing a local computer that they have physical access to [33]. As suggested by the table above, when using a local computer, CFIs have maximum access which facilitates the examination and analysis of a local computer.

Table 1 Different types of information provided using different service models

Information	SaaS	PaaS	IaaS	Local
Networking	✗	✗	✗	✓
Storage	✗	✗	✗	✓
Servers	✗	✗	✗	✓
Virtualization	✗	✗	✗	✓
OS	✗	✗	✓	✓
Middleware	✗	✗	✓	✓
Runtime	✗	✗	✓	✓
Data	✗	✓	✓	✓
Application	✗	✓	✓	✓
Access control	✓	✓	✓	✓

However, the major problem with cloud services that a CFI often faces is that the evidence needs to be provided by the CSP. Hence, if CFIs require application logs, database logs, or network logs when utilizing the SaaS service model, they will need to contact the CSP so as to acquire it since they cannot access it by themselves [18, 33]. Another major problem with cloud services that a CFI often faces is that the user's data is maintained with the data of other users on the same storage systems (see also Sect. 4 for other challenges to cloud services). This culminates in difficulties in segregating only the data that CFIs require to carry out the examination and analysis process. For example, if two users are using the same webs server to host a web page, it would be difficult to prove that the server's log contains the data of the user that CFIs are looking for. Thus, this will be an issue when conducting a forensic analysis of the cloud service. In the following list we discuss every entry in the table above [10, 18, 29, 33]:

- Networking: In a local environment, investigators have access to the network machines, such as switches, routers, IDS/IPS systems, etc. Hence, they can access all of the traffic transmitted through the network, examine and analyze it to acquire as much data as possible. However, when using the cloud, even the CSPs do not have that kind of data as described as they must not log all the traffic transmitted through the network, since users' data is confidential and CSP are not permitted to record, store, and analyze it.
- Storage: When investigators have physical access to the machine, they know where the data is store. However, when using a cloud service, the data could be located in various jurisdictions.
- Servers: In a local environment, investigators have physical access to the machine to analyze the data on it (all the data is stored on the machine). However, using the cloud, this is not possible as the data is spread over several data centers, and it is difficult to confirm that they have acquired all the data required.

- Virtualization: In a local environment, CFIs have access to the virtualization environment, where they can access the hypervisor, manage existing virtual machines, delete a virtual machine, or create a new virtual machine. In the public cloud, often they do not have access to the hypervisor. However, access can be obtained in private clouds.
- OS: In a local environment, investigators have complete access to the operating system as they have in the IaaS model, but not in the PaaS and SaaS models. If the investigators need to access the operating system, they must connect to the SSH service running on the server and issue OS commands.
- Middleware: The middleware connects two separate endpoints, which together constitute a complete application.
- Runtime: Investigators can determine how the application must be initiated and stopped when using the IaaS model so that they can have access to its runtime.
- Data/application: In PaaS and IaaS models, investigators can access all data and applications, which we can manage by using search, delete, add, etc. However, they cannot do this directly when using the SaaS model.
- Access control: In all service models, investigators have access to the access control since without it they will not be able to access the service. Investigators can control how access is granted to different users of the application.

It is important to note that when performing a cloud network forensic examination and analysis, investigators do not have access to the same information as they have when carrying out an examination and analysis of a local computer system. Investigators often do not have access to the information that they are searching for and must request the CSP to provide the information they require. One of the issues with such an approach is that the investigators must rely upon the CSP to grant them the appropriate information. CSPs might decide to provide false information or hold back some vital information. This is a major issue in particular in circumstances where CFIs are to use the data in court, since they will need to prove without a doubt that the evidence from the collected data belongs to the user. The process of identifying, preserving, acquiring examining, analyzing, and interpreting the data must be documented through chain of custody, and must be admissible in courts [7, 21]. When an attack is carried out over a cloud service, CFIs will need to deal with several different problems. However, the most vital problem is interaction with the CSP. Due to the fact that the services reside in the cloud, there is a large volume of information that could serve as digital evidence that can only be provided by CSP as it is the CSP that has access to it [18]. There exist also other issues with collecting the data when dealing with cloud environments, such as when data is stored in multiple data centers located in various different jurisdictions, or when data of different users are stored in the same storage device, etc. [10, 29]. Therefore, there is still a lot of research that must be carried out in order to enhance digital forensic investigations of cloud services. Moreover, there still exists a lack of qualified cloud forensic investigators. Therefore, the number of cloud forensic professional will also need to increase in the near future.

4 Challenges to Cloud Forensics

As of now, establishing forensic capabilities for cloud organizations in the three dimensions that were defined earlier in the paper will be difficult without overcoming several vast challenges. For instance, the legal aspect currently does not have agreements among cloud organizations in relation to collaborative investigation. Moreover, most of SLAs do not have terms and conditions available with regards to segregating responsibilities between the CSP and customer. Policies and cyber laws from various jurisdictions must also perform their part so as to solve conflicts and issues resulting from multi-jurisdictional investigations. This section presents eight challenges that prevents the establishment of a cloud forensic capability that encompass the 'technical', 'organizational' and 'legal dimensions'.

4.1 Challenges to Forensic Data Collection

In all circumstances implicating cloud service and deployment models, the cloud customer encounters issues in relation to decreased access to forensic data based on the cloud model that is implemented [29]. For instance, IaaS users might enjoy relatively easy access to all data needed for forensic investigation, whereas SaaS customers might have little or no access to such data [10]. Lack of access to forensic data denotes that the cloud customers will have little control (or no control) or even knowledge of where their data is physically located. Cloud customers might only be able to specify the location of their data at a higher level of abstraction, typically as a virtual object container. This is due to the fact that CLSs deliberately hide data locations the actual location of the data in order to assist data movement and replication [18]. Moreover, there exists a lack of the terms of use in the Service Level Agreements in order to facilitate forensic readiness in the cloud. Many CSP purposely avoid offering services or interfaces that will assist customers in collecting forensic data in the cloud. For example, SaaS providers do not provide IP logs or clients accessing content, while IaaS providers do provide copies of recent Virtual Machine states and disk images. The cloud as it operates now (as of 2016) does not offer customers with access to all the relevant log files and metadata, and limits their ability to audit the operations of the network utilized by their provider and conduct real-time monitoring on their own networks.

4.2 Challenges to Static and Live Forensics

The propagation of endpoint, particularly mobile endpoints, is one of the major challenges for data discovery and evidence acquisition. The large number of resources connected to the cloud makes the impact of crimes and the workload of investigation even larger [29]. Constructing the timeline of an event needs accurate time synchronization which is vital in relation to the audit logs employed as source of evidence in the investigations [10]. Accurate time synchronization is one of the major issues during network forensics; and it is often aggravated by the fact that a cloud environment needs to synchronize timestamps that is in harmony with different devices within different time zones, between equipment, and remote web clients that include numerous end points. The usage of disparate log formats is already an issue in traditional network forensics. The issue is aggravated in the cloud because of the large volume of data logs and the pervasiveness of proprietary log formats. Analogous to other branches of forensics, deleted data in the cloud is considered as a vital piece of artifact. In the cloud, the customer who created a data volume often maintains the right to modify and remove the data. When the customer removes a data item, the deletion of the mapping in the domain begins immediately and is typically completed in seconds [10, 29]. After that, there is no way to access the removed data remotely, and the storage space, having been occupied by said data, becomes available for future write operations, and it is possible that the storage space will be overwritten by newly stored data. However, some removed data might still be present in a memory snapshot. Therefore, the challenge is to recover the deleted data, identify the ownership of the deleted data, and employ the deleted data for event reconstruction in the cloud.

4.3   Challenges in Evidence Segregation

In a cloud environment, the various instances of virtual machines running on the same physical machine are isolated from each other via virtualization. The instances are treated as if they were on separate physical hosts, and as such, they will have no access to each other despite being hosted on the same machine [5]. Customer instances do not have access to raw disk devices, instead they have access to virtualized disks. Technologies employed for provisioning and deprovisioning resources are constantly being updated [10, 5]. CSPs and law enforcement agencies often face a challenge to segregate resources during investigations without violating the confidentiality of other tenants that share the same physical hardware, while also ensuring the admissibility of the evidence [18]. Another challenge is that the easy-to-use feature of cloud models facilitates a weak registration system. This makes anonymity easier which enables cybercriminals to hide their identities and more difficult for investigators to detect and trace perpetrators. CSPs employ encryption in order to segregate data between cloud customers. However, when this feature is not available, customers are often encouraged to encrypt their sensitive data before uploading it to the cloud [29]. Therefore, it is suggested that the segregation must be standardized in SLAs and access to cryptographic keys must also be formalized consistent with CSPs, consumers and law enforcement agencies.

4.4   Challenges in Virtualized Environments

Virtualization is a significant technology that is employed to implement cloud services. However, virtualised environments pose several challenges. Cloud computing provides data and computing power redundancy by duplicating and distributing resources. Many CSPs do this by employing different instances of a cloud computer environment within a virtualized environment, with each instance running as a stand-alone virtual machine that is monitored and maintained by a hypervisor [10]. This denotes that attackers can target the hypervisor, and doing so successfully provides them with free control over all the machines being managed by it. However, at the same time, there exists a lack of policies, techniques, and procedures on the hypervisor level that could assist CFIs in conducting cloud forensic investigations. Another challenge presented is the loss of data control. Therefore, tools and procedures must be developed in order to identify forensic data physically with specific timestamps while at the same time considering the jurisdictional issues. Digital forensic readiness—or proactive measures which include both operational and infrastructural readiness—can significantly assist cloud forensic investigations. Examples include, but are not limited to, preserving regular snapshots of storage, continually tracking authentication and access control, and performing object-level auditing of all accesses.

Cloud computing provides data and computing power redundancy by duplicating and distributing resources. Most CSPs implement redundancy employing various instances of a cloud computer environment within a virtualized environment, with each instance running as a stand-alone virtual machine that is monitored and maintained by a hypervisor. A hypervisor is similar to a kernel in a traditional operating system. This denotes that attackers can target the hypervisor, and doing so successfully provides them with control over all the machines being managed by it. Data mirroring over multiple machines in various jurisdictions and the lack of clear, real-time information about data locations presents challenges in forensic investigations [8]. Moreover, a CSP cannot offer an exact physical location for a piece of data across all the geographical regions of the cloud. Also, the distributed nature of cloud computing necessitates robust international cooperation, particularly when the cloud resources to be seized are located around the world [18, 29].

4.5   Challenges in Internal Staffing

Currently, most cloud organizations are only dealing with investigations employing conventional network forensic tools and staffing, simply neglecting the issue [10]. The major challenge in establishing an organizational structure in cloud forensics is finding the right amount of expertise and relevant legal experience within the available manpower. The major challenge is presented by the scarcity of technical and legal expertise in relation to cloud forensics. This is worsened by the fact that forensic research and laws and regulations lag behind the rapidly-evolving cloud technologies [3]. Therefore, cloud companies must ensure that they have adequate qualified and trained staff to deal with the technical and legal challenges existing in cloud forensic investigations.

4.6 Challenges in External Chain of Dependencies

CSPs and most cloud-related applications have dependencies on other CSPs. This denoted that investigation in the chain of dependencies between CSPs will depend on the investigations of each link in the dependency chain, which means that any corruption or interruption between all the parties involved can result in major problems for the entire investigation. There are currently no tools, policies, procedures, or agreements that deal with cross-provider investigations [10, 29].

4.7 Challenges in Service Level Agreements

Because of the absence of customer awareness, the lack of CSP transparency and international regulations, the majority of cloud customers end up not having awareness of what has occurred in the cloud in cases in which their data is lost or compromised because of criminal activity.

4.8 Challenges to Multi-jurisdiction and Multi-tenancy

The presence of multiple jurisdictions and multi-tenancy in cloud computing present several significant challenges to forensic investigations. Each jurisdiction enforces dissimilar requirements in relation to data access and retrieval, evidence recovery without violating tenant rights, evidence admissibility and chain of custody. The lack of a global regulatory body significantly affects the effectiveness of cloud forensic investigations.

5 Opportunities

Although there exist many challenges involved in cloud forensics, at the same time there are various opportunities that can be taken advantage of in order to advance forensic investigations. Six opportunities are introduced in the following sections:

- *Cost Effectiveness*: Similar to the cloud technology, that makes things less expensive when implemented on a larger scale, cloud forensic services and security will be more cost effective when implemented on a global scale. This denotes that companies that cannot afford dedicated internal or external forensic capabilities might be able to capitalize on low-cost cloud forensic services.
- *Data Abundance*: Because of providers' practice of maintaining redundant copies of data to ensure robustness, cloud forensic investigators will be able to capitalize on the abundance of data, particularly because the redundancy makes it more difficult to delete data and enhances the probability of recovering deleted data.
- *Forensics as a Service*: The concept of FaaS is gradually developing in cloud computing and demonstrating the advantages of a cloud platform for large-scale digital forensics [27, 23]. Security vendors are altering their delivery methods in order to cover cloud services, and some companies are providing security as a cloud service [29]. In the same way, forensics as a cloud service could capitalize upon the enormous computing power of the cloud to support cybercrime investigations at all levels.
- *Establishment of Standards and Policies*: Forensic policies and standards are often behind technological advancements, culminating ad hoc solutions [20, 30]. With regards to technology, forensics are often treated as "afterthoughts and bandage solutions", only being created after the technologies have matured [10]. However, with cloud computing, there exists an opportunity to establish policies and standards while the technology is still in its infancy.

6 Conclusions

The swift developments and increase in popularity of cloud technology is certainly orienting the domain of digital forensics to a whole new level. Cloud computing also poses new questions in relation to who owns the data and also with regards to the customers' expectations of privacy. Many existing challenges may be aggravated by the cloud technology, such as various jurisdictional issues and lack of international harmonization and coordination. Laws vary on the legal protections related to data in the cloud from country to country. However, the cloud environment also presents unique opportunities that can significantly advance the efficacy and speed of forensic investigations and provide a ground for novel investigative approaches. As future work, a list of comparisons about privacy and how to deal with confidential can be developed.

References

1. ACPO. (2012). ACPO Good Practice Guide for Digital Evidence, *U.K. Association of Chief Police Officers.*
2. Beebe, N., & Clark, J. (2005). A hierarchical, objectives-based framework for the digital investigations process. *Digital Investigation, 2*(2), 147–167.
3. Beebe, N. (2009). Digital forensic research: The good, the bad and the unaddressed. In *International Conference on Digital Forensics* (pp. 17–36). Berlin: Springer.
4. Broadhurst, R. (2006). Developments in the global law enforcement of cybercrime. *Policing: International Journal of Police Strategies and Management, 29*(2), 408–433.
5. CSA. (2009). Security Guidance for Critical Areas of Focus in Cloud Computing V2.1. Retrieved October 11, 2016, from https://cloudsecurityalliance.org/csaguide.pdf.
6. Carrier, B., & Spafford, E. (2003). Getting physical with the digital investigation process. *International Journal of Digital Evidence, 2*(2), 1–20.
7. Casey, E. (2011). *Digital evidence and computer crime: Forensic science, computers and the internet* (3rd ed.). New York: Elsevier Academic Press.
8. Catteddu, D. (2010). Cloud computing: Benefits, risks and recommendations for information security. In *Web application security* (pp. 17–17). Berlin: Springer.
9. Chabrow, E. (2011). 5 Essential Characteristics of Cloud Computing. Retrieved October 10, 2016, from http://www.inforisktoday.co.uk/5-essential-characteristics-cloud-computing-a-4189.
10. Cruz, X. (2010). The Basics of Cloud Forensics. Retrieved October 10, 2016, from http://cloudtimes.org/2012/11/05/the-basics-of-cloud-forensics/.
11. EurActiv. (2011). Cloud computing: A legal maze for Europe. Retrieved October 10, 2016, from http://www.euractiv.com/section/innovation-industry/linksdossier/cloud-computing-a-legal-maze-for-europe/.
12. Freiling, C., & Schwittay, B. (2007). A common process model for incident response and computer forensics. In *3rd International Conference on IT-Incident Management & IT-Forensics,* (pp. 19–40).
13. ISO/IEC. (2011). *ISO/IEC 27035. Information technology–security techniques–information security incident management.* Geneva, Switzerland: International Organization for Standardization.
14. ISO/IEC. (2015). *ISO/IEC 27043: Incident investigation principles and processes.* London: British Standards Institution.
15. Kent, K., Chevalier, S., Grance, T., & Dang, H. (2006). Guide to integrating forensic techniques into incident response. *NIST Special Publication,* 800–86.
16. Kohn, M., Eloff, M., & Eloff, J. (2013). Integrated digital forensic process model. *Computers & Security, 38,* 103–115.
17. Liles, S., Rogers, M., & Hoebich, M. (2009). A survey of the legal issues facing digital forensic experts. In *International Conference on Digital Forensics* (pp. 267–276). Berlin: Springer.
18. Lukan, D. (2014). Cloud Forensics: An Overview. Retrieved October 11, 2016, from http://resources.infosecinstitute.com/overview-cloud-forensics/.
19. Mell, P., & Grance, T. (2011). The NIST definition of cloud computing. *Communications of the ACM, 53*(6), 50.
20. Meyers, M., & Rogers, M. (2004). Computer forensics: The need for standardization and certification. *International Journal of Digital Evidence, 3*(2), 1–11.
21. Montasari, R. (2016). A comprehensive digital forensic investigation process model. *International Journal of Electronic Security and Digital Forensics (IJESDF), 8*(4), 285–301.
22. Montasari, R., & Peltola, P. (2015) Computer forensic analysis of private browsing modes. In *Proceedings of 10th International Conference on Global Security, Safety and Sustainability:*

Tomorrow's Challenges of Cyber Security (pp. 96–109). London: Springer International Publishing.

23. Oberheide, J., Cooke, E., & Jahanian, V. (2008). CloudAV: N-version antivirus in the network cloud. In *Proceedings of the Seventeenth USENIX Security Conference* (pp. 91–106).

24. Palmer, G. (2001). A road map for digital forensic research. 1st Digital Forensic Research Workshop (DFRWS) (pp. 27–30).

25. Perry, R., Hatcher, E., Mahowald, R., & Hendrick, S. (2009). Force. com Cloud Platform Drives Huge Time to Market and Cost Savings. *IDC White Paper*, International Data Corporation, Framingham, Massachusetts.

26. Quick, D., & Choo, K. (2016). Big forensic data reduction: Digital forensic images and electronic evidence. *Cluster Computing*, 1–18.

27. Roussev, V., Wang, L., Richard, G., & Marziale, L. (2009). A cloud computing platform for large-scale forensic computing. In *International Conference on Digital Forensics* (pp. 201–214). Berlin: Springer.

28. Rowlingson, R. (2004). A ten step process for forensic readiness. *International Journal of Digital Evidence, 2*(3), 1–28.

29. Ruan, K., Carthy, J., Kechadi, T., & Crosbie, M. (2011). Cloud forensics. In *International Conference on Digital Forensics* (pp. 35–46). Berlin: Springer.

30. US-CERT. (2012). Computer Forensics. U.S. Department of Homeland Security. Retrieved June 17, 2006, from https://www.us-cert.gov/security-publications/computer-forensics.

31. Valjarevic, A., & Venter, H. (2015). A comprehensive and harmonized digital forensic investigation process model. *Journal of Forensic Sciences, 60*(6), 1467–1483.

32. Wilson, D. (2015). Legal Issues with Cloud Forensics. Retrieved October 12, 2016, from http://digital.forensicmag.com/forensics/april_may_2015?pg=18#pg18.

33. Zawoad, S., & Hasan, R. (2013). Digital Forensics in the Cloud, In Alabama University in Birmingham (pp. 1–4).

Business Intelligence Tools for Informed Decision-Making: An Overview

Abimbola T. Alade

Abstract Numerous organisations are facing challenges of manipulating large volumes of data generated as a result of their internal business processes. Manual inference from such data often results in poor outcome. Decision makers within such firms are now reliant on the processed data created by the use of business intelligence tools and dashboards. In this chapter, the business intelligence and analytics concepts are explained as a means to manage vast amounts of data. Number of business intelligence tools and relevant strategies are discussed. Case studies and applications, e.g., banking sector are provided.

Keywords Big data · Strategy · Business intelligence

1 Introduction

The term big data is now no longer as mysterious as it used to be when it first became 'a thing' in the 1990s. The most prominent discussions surrounding big data of late, is how to successfully harness the treasure it holds.

Strategy according to [1] is how to win. It is a guide, a prescribed set of actions to achieve organisational goals.

Strategic big data therefore is a process, a set of actions that can help organisations win using big data as an asset.

Big data is 'data sets so large and diverse', that they defy the regular or common IT setups [2]. Elias further stated: agreeing with the opening paragraph of this section that, defining big data is not near as important as knowing what

A.T. Alade (✉)
School of Computing, Creative Technologies & Engineering, Leeds Beckett University, Leeds, UK
e-mail: a.alade2242@student.leedsbeckett.ac.uk

© Springer International Publishing AG 2017
A. Hosseinian-Far et al. (eds.), *Strategic Engineering for Cloud Computing and Big Data Analytics*, DOI 10.1007/978-3-319-52491-7_12

organisations can do by applying analytics to big data which is what the rest of this chapter tries to detail.

2 Big Data

The term big data as the name implies is simply huge amounts of data, which can be described in terms of volume, velocity and variety. Volume describes the vast amounts of available data, velocity describes how fast the available data increases and variety describes the various forms these data exist—structured or unstructured or the ability of companies to mine information of significance from large volumes of data, which requires investing in tools, methods and control [2, 3].

Big data is created from emails, social data, financial transactions, GPS and satellite images, delivery or pickup services like taxis, etc., access to users tagged locations as they update their profile or upload pictures, video or audio files, or as they use the 'check-in' function on Facebook, which basically records and make visible to friends on the platform, their location at the point in time. Big data also comes from data generated using mobile phones and the web, etc. [4]. The works of [5] confirm the aforementioned and also prove that big unstructured data can be sourced from social media by the process of 'topic modelling' which confers competitive advantage on companies, though [6] pointed out a limitation to this method of data mining, which is the fact that its effectiveness does not extend to searching text strings or phrases—Variety.

Big data is what generates data for analytics. It is unstructured data that cannot be stored directly into a structured database for analysis without prior processing. It means enormous quantity of unstructured and partly structured data from different sources can disclose a number of consistencies in arrays of data and also makes irregularities in data known, provided that similarity in patterns exist in the data set being examined.

The rate at which data is growing rapidly—velocity—makes more businesses desire to explore the potentials data holds for them, more than ever [7].

As data keeps increasing in volume so does the need for faster ways of processing and analysing data and increase data-driven decision-making. Also, in order not to get confused and overwhelmed with the amount of big data and its sources Hadoop as a tool can be used to store vast amounts of data, carry out Extraction, Transformation and Loading of data into a structured database in order for these data to be in an acceptable form, from which meaningful analysis can be made [8, 9].

In order to efficiently and strategically manage big data, this chapter would go a step further by suggesting an approach to big data management—Business intelligence.

3   Business Intelligence

Business Intelligence—BI is the 'umbrella term' used for the design of systems, tools, databases and methods that is used to make organisations smarter, including all activities that involves and supports collecting, storing, processing and analysing data that is crucial to supporting business decisions [10, 6].

BI is also known as Management information system, Executive information system, Information Advantage, Knowledge management and Online Analytical processing according to the studies of [11–13]. BI is a tool or set of concepts used in Analysis, which provides answers to unstructured business queries and automates business decisions [14]. Such that needs to be made from several sources using historical data to understand the state of a business per time, in order to guide future actions and also predict future occurrences, these decisions also involve carrying the right people along with the right sets of information at the right time.

Ghazanfari et al. [11] also raised two main schools of thoughts as it relates to BI, which they called the Managerial and Technical approach to BI, also referred to as the two core roles of BI. To them, the managerial role of BI is the process of analysing information collected from within and outside the business in order to gain useful insights for the organisations strategic decision-making. Which is the same as what was described by [9] that BI uses technology to bring information together in any form for analysis, information which could lie internally or externally and could be in any form, so as to intensify organisational performance. The technical approach on the other hand refers to BI as the tools that support the managerial process of gathering and analysing information. The managerial approach to BI is what this write-up is proposing to organisations as a way of harnessing and strategically managing big data, while making use of the technical approach of tools as well.

The objective of BI is to improve timeliness, meaningfulness and the worth of every factor that goes into decision-making. BI tries to create smart business, by giving the right people, who have asked the right questions and the right information [10].

This writing focuses on both core aspects of BI in terms of the tools suitable for managing big data for organisations and the knowledge that is derivable from the use of these tools.

Now, the fact that the gathered information can be massive and can be measured in Exabyte's' gave rise to what is known as Big data, requiring hi-tech tools for analysing the hidden knowledge that can be sieved out of data sets and used to address complex organisational issues.

Latest trend in BI is that practitioners are being more data aware and they are paying attention to the management or 'health of their data' knowing that good quality data would yield quality analysis.

Practitioners are also beginning to be data aware to the point of seeking the tools they can work with within their organisation. That is, data analytics is becoming part of what organisations observe as a core personalised task and taking active steps towards, wanting more, to the point of setting up COE—Centre of Excellence for analytics—having standalone mobile analytics, investigating data outside of their primary domain, thereby creating the need for more BI tools [15].

This is not to say that the introduction of BI to big data authorises discarding corporate data [4] or is a 100% path to success. [16] points out that, for BI projects to be successful, the data being analysed has to be of good quality. There also has to be an understanding between IT staff and business users. There is need to train business users enough to be comfortable with the new system. There has to be clarity of the fact that the use of BI is not a quick fix, time efficiency in BI is relative to the complexity and level of analytics required and lastly, introduction of BI to big data is not the licence to drop the old tools but the trick is coordinating both to get the best business results.

A Study by [17] shows that the concept of knowledge management which is also a synonym for BI, is not sector specific, neither does it stand alone, in the sense that there has to be a collaboration between people, process and technologies to have successful knowledge management. Which definitely grants 'competitive differentiations' amongst peers in the same sector. The study further shows that possessing knowledge by corporations is not enough as an end but the ability to leverage on these knowledge is in itself strategy, deployed by organisations.

BI involves building the required systems and being able to use it for analytics and reporting among other things as stated by [18], which is in agreement with the works of [10, 6].

The benefits of BI in summary is that, it helps understand market trends, customers wants and competitors' actions from within a variety of voluminous data that keeps increasing. Business intelligence helps to bring data together from different sources to aid analyses, improves business performance, simplifies decision-making, creates a single view of organisational information, makes data easily sharable and accessible at any time, allows faster and more accurate and higher quality reporting and increases revenue [19, 12].

BI consists of up to four aspects or components according to [9], as introduced late in the 2000s [20], which will be discussed in the next section of this writing.

Business Intelligence in itself is as old as the 1990s when it became popular in the world of Business and Information technology (IT) with OLAP being the technology behind a lot of BI applications or characteristics, used to discover data including complex analytical calculations, reporting and even in predicting outcomes. In their study on the future of BI, [18] linked Agile to BI, while evaluating Agile principles. This writing finds the second agile principle most relevant, which talks about valuing working software over comprehensive documentation, which validates one of the common themes of this writing, that knowing what big data is or the amount an organisation possesses is not as important as knowing that can be derived from it and in real time.

According to [18] 'fast analytics' is a phenomenon to be dealt with in this age of big data, due to the vast amounts of available data.

3.1 Business Intelligence Components

Data warehouse and Business Process Management were two of the main BI components highlighted by [12] in their study. Others are Data mining, Data marts, Decision support and Management support systems. Four main components of BI will be discussed in this section as proposed by [21].

3.1.1 Data Warehouse

Data warehouses are repositories of data, which is as current as at the most recent time the warehouse is loaded with transactional data from various sources. Data loaded into the data warehouse becomes historical data from Online Transactional Processing (OLTP) systems.

OLTPs can also provide analytic insights and reporting functions. However, businesses may not be able to rely on these analyses, as they are resultant from transient data that keeps changing daily. As opposed to insights derived from Online Analytical Processing (OLAP), systems that are historical, loaded as at a particular date and are capable of analysing and presenting information in an easily comprehensible form, which bestows competitive advantage over industry peers and makes apt decisions possible for organisations. Data warehouses make it possible to access data from different sources, and easily produce knowledge. The data stored in the data warehouse would have gone through an ETL process, i.e., would have been extracted from the OLTP systems, transformed to remove inconsistencies and protect data integrity before being loaded into the warehouse, where reporting is done with data mining and OLAP tools [6, 22].

3.1.2 Business Performance Management (BPM)

The studies of [23, 16, 24, 25]; Web Finance Inc [26] posits that BPM, also known as Enterprise or Corporate Performance Management is a set of approach and tools used in business management, which takes account of all the activities that form part of the business as a whole, aligns the objectives of these activities, reviews them and delivers the facts about the business standings to the management, as well as suggesting how to meet these outstanding goals in order to arrive at decisions that can improve the overall performance of the organisation.

BPM makes use of reports, analytics and tools such as 'Bonita soft BPM', that graphically represents business processes with signs and symbols in order to secure the same level of understanding within an organisation so as to enhance performance, which can occur when everyone is aware of the available levels of activities involved in a task and the order of those activities as well as whose responsibility it is to carry it out as it is the intersect between business and information technology teams.

BPM also uses ideologies such as TQM—Total quality management, six sigma and lean, which are agile project management approaches that believe in trimming off irrelevances within a process.

For BPM to be effective, organisations have to acknowledge the external environment.

3.1.3 User Interface

User interfaces can be graphical, in text, or audio form and makes it possible to access information easily from BI systems by providing one point of interface between business users and the BI systems/solutions that processes the information users require, and returns information or options that can aid decision-making after eliminating below par alternatives. This interaction between man and machine reduces human inputs but maximises relevant machine outputs. User interfaces can be dashboards in the form of graphs and charts which summarises important information and can also be in form of Google maps, satellite navigation systems that are Geographic information systems—GIS, operating or input control buttons or devices, computer mouse, Apples Siri, Cortana, search engines, mobile apps, etc. [22, 27–29].

3.1.4 Business Analytics and Visualisation

Business Analytics is the skill techniques and process of using purpose built tools to explore a set of data to find relationships between them while concentrating on the statistical implications of such data. As well as being able to explain it so as to support decision-making guided by facts within data, with the ultimate aim of being at an advantage. Data visualisation helps to find abnormalities and outliers' resident within a data set, and helps in identifying patterns worthy of further probing, to aid data understanding [30].

Business analytics aids the automation process in businesses, test past decisions and can predict future consequences—also known as predictive modelling or analytics, which SAS enterprise miner, R, Python and other tools help carry out.

The predictive power of analytics helps to know what the result of each course of action will be. Such as if sales are going to be made, if crime will be committed—when and where and if there will be default on payments for goods and services among other things, based on past occurrences evidence of which is available in gathering and accessing historic data and test models that are built to assess new models.

Business Analytics helps answer questions like why something happened, if it will happen again, the impact of actions taken and hints on expositions that cannot be seen by merely looking at or 'eye balling' data [31, 32].

Organisations in all sectors in advanced countries use analytics to class customers according to their risk category in order to understand their needs and measure their exposure to customers in different classes.

Visualisation is the display of important, complex and nonconcrete information, buried within data with the use of visual aids such as graphs, charts, maps, figures, tables, etc., in order to give it shape and form making it possible to describe a situation/occurrence in a way that the intended can easily comprehend and value [16, 9]. Data visualisation helps summarise bulky data.

The combination of both concepts described above is what this writing is about with the intention of suggesting ways to exploring big data and bringing to fore, the advantages therein.

The question may arise as to where the aforementioned historical data exists or would come from, the answer is that, data is generated with every instance of the use of technology, which about 85% of the global population have continued to supply, by transacting using debit/credit cards or cash, participating in online purchases, use of bus passes, oyster cards, online activities, using the 'check-in' button on Facebook, search histories, purchase histories, anything for which a record can be created or logged, these among other things contribute to the vast amount of big data available for exploiting. Data has since become a commodity that companies buy and sell in order to carry out target marketing campaigns, know more about their cusAs an analytics tooltomers, etc., through analytics even much more than the customers will like to give away, according to gartner group CIO survey of 2013 [33] data analytics and BI rank as major areas of investment focus.

4 Business Intelligence Tools

Business Intelligence tools refer to software's or technologies that aid decision-making within organisations. They support businesses in gathering, organising, analysing, understanding and converting data into a form that helps organisations arrive at decisions that are more precise.

Several BI tools have been in existence over the years and more continue to be created as needs or requirements surface in the industry requiring them [34]. A few of the tools obtainable in the BI tools market today are Sisense, Dundas, Izenda, Zoho reports, Inetsoft, Tap Analytics, Adaptive discovery, Pc Financials, etc., according to [35]. The following were also mentioned in [36] as top BI tools while sighting [37] Oracle BI, Pentaho, Profit base, Rapid insight, SAP BI, SAP Business Objects, Silvon, Solver, SpagoBI, Style intelligence, Targit, Vismatica, WebFO-CUS, Yellowfin BI, Ducen, Clear Analytics, Bizzscore, Good data, IBM cognos intelligence, Insightsquared, Jaspersoft, Looker, Microsoft BI, Microstrategy, MITS, OpenI, among many others.

Sherman [34] classified BI tools into three categories according to the depth of analytics they perform.

Guided analysis and reporting: these are tools that have been used through the years to perform different analysis with the same sets of data, this category of tools was mainly used to create static reports, but a bit of dynamism is now being introduced into the reports generated from these sets of tools, as users are now able to compare data in different time periods, visualise data and analyse data with a combination of other tools. These tools are used to create static and dynamic reports, scorecards and dashboards among others.

Self-service BI and analysis: these include tools used to perform impromptu data analysis of a mathematical or statistical nature, where reports mostly one-off are needed by the top management, in situations where the task to be completed is almost always more than the amount of time allowed to complete the tasks. These tools enable users to be able to add new data apart from what is obtainable within the business or what the business has always used, and can be used to discover and visualise new data, the three BI tools (SAS, Qlikview and Tableau) being examined in this work fall in this category.

Advanced analytics: these sets of tools are used by data scientists to perform tasks that are predictive and prescriptive in nature, which is a level above what data analysts or what data reporters do. These tools make use of available data to prescribe an appropriate line of action or decision for the business or predict likely outcomes which are different from the descriptive nature of the tools used in the two aforementioned categories above.

Furthermore, acknowledging the fact that the list of existing BI tools is not limited to those stated above, the tools—Qlikview, SAS and Tableau—examined in this work are suitable for most organisations, are highly affordable, compatible with windows system, which most organisations use, so the issue of investing heavily in new systems is not something to worry about. The tools are all deployable on site, while deploying remotely via the cloud is available to Qlikview and SAS.

In terms of sophistication, all the tools mentioned above are all rated with four stars and above ratings, they also have the advantage of ease of use in the sense that they can be used by individual users with Qlikview having the edge of up to 51 users per time. They all also have recommendations by no less than 83% but no more than 94% of users.

It is important to reiterate at this point that it is not enough to put a BI tool to use or have it ready for use, a harmonious relationship between the business users and IT staff of an organisation is invaluable, and somewhere in between should be far-reaching and correct data in order to get the benefits these tools promise for organisations, even banks. Especially as [38] posits that BI tools can help banks combat problems they face, which [39] states could be that of analysing who to lend to and who could have repayment challenges at a future date.

5   The BI Tool SAS

The Business intelligence tool SAS has been judged suitable for use in the many sectors and had been around for 30 years at the time of [40] writing, this year makes it 39 which is in correlation with [20] in mentioning BI and analytics as being founded on statistical methods developed in the 70s.

Despite the fact that it is a specialised analytics software, SAS is usable by anyone as long as such person knows the business question and the required data, in addition to the explanation of the generated results. SAS is a part of day-to-day living, from knowing the best course of action for an endeavour, knowing the right marketing audience, to price setting for businesses, etc.

SAS is also usable on many platforms such as the Internet, desktop computers, personal computers, Microsoft excel and PowerPoint applications, which makes it possible to analyse unprocessed data more easily.

SAS used to mean Statistical Analysis Systems, but now it is the name of the company that develops SAS software's such as SAS enterprise guide, SAS Base, SAS enterprise miner, etc. It also used to mean statistical analytics software. SAS serves as the answer to many questions in several industries and countries all over the world.

Thus, it can be used to analyse data coming from any source, be it relational databases, such as Oracle, SQL Server, DB2 or personal computer data sources such as Microsoft excel, access, text documents or CSV—comma separated value files. SAS server receives these data and the analysed, summarised and reported output are temporary business needs.

However, SAS can go further by giving organisations the benefit of high level, more permanent results, and long-term approach by making the best use of available information and opportunities while saving time and money by making it possible for existing reports to be updated when new data is uploaded, which can easily be accessed by the business units requiring it. SAS also performs predictive functions in a business sense.

The following are a few of the statistical techniques SAS uses as they relate to different fields and analysis:

- Survival analysis: to predict the average time before there is a downturn of a variable, e.g., the average time before sales fall, etc., based on test models.
- Forecasting: to know the impact of a variable on another at a particular time.
- Mixed models: to know the impact of doubling a variable on another.
- Data mining: to know the projected profit from acquiring a new customer based on their demography, e.g., their age, income, number of children, etc.
- Categorical data analysis: to know the difference between two or more categories of variables.

As earlier mentioned, there is a relationship between Microsoft, BI tools and MS excel; which in itself is a ready tool for BI and one of the most 'common' and often overlooked Microsoft tool.

With or without the SAS add-ins, which is like a SAS component that is added to the right version of Microsoft excel for both to be used together, pivot tables are desktop OLAP tools which make it easy to summarise and explore large data sets.

Organisations can benefit from creating dynamic dashboards using SAS add-ins in Excel. These dashboards can show the performance of businesses and classification per period of time, by drilling up to the figures of the organisation as a whole and drilling down through regions, branches, accounts and customers or better still multidimensional analysis which can make it possible for any deterioration in a product, branch or regional performance to be spotted easily.

SAS also makes it possible for data to be stored in and accessed from a central place through the information delivery portal and SAS web report studios which is a benefit to businesses who have their unit staff positioned in different locations, thereby sharing results [40] and everyone within that business unit 'knows what everyone knows' [14].

Over the years, SAS has continued to be a tool of high value to the success of many endeavours, for example, in the just concluded Rio Olympics; SAS was the analytics tool employed to deliver gold to Great Britain's rowing team.

In addition, diverse institutions including many banks across the world are SAS testifiers, for example.

Bank of America makes use of SAS to reduce processing times, make quick decisions and be in the forefront of the market. Bank of India gets the benefit of a consolidated view of data from seven different portfolio positions in 17 currencies in a day, therefore bestowing the benefit of running or compiling risk reports daily, which the bank compiled quarterly, before SAS. Barclays bank is enjoying the benefits of SAS BI dashboard with which they can now monitor individual performances as it affects overall performance. South Africa's Absa also testifies to gaining competitive position with SAS. Australia's Bendigo and Adelaide Bank have gained new insights into risk in time and without needing to worry about accuracy, a benefit RSA Canada also enjoys with the use of the SAS software. Brazil's Cemig also uses SAS to combat fraud [41].

6 The BI Tool Tableau

Tableau allows whoever is analysing data to be able to create their own visualisation manually or get recommendation as required based on the data they are working with [30]. This tool on its own makes data visible, tableau makes visual analysis possible with charts, graphs, maps, etc. Computer graphics introduced data visualisation in the 50s, and it is the science of making data discernible [10] more precisely to 'see the unseen'. By importing data into the software, better known as 'connecting to data' from spreadsheets—usually excel, it is possible to view the same data within tableau as was in the spreadsheet, by clicking on the data sheets listed on the left side of the connection canvas and preview pane, which is displayed in the preview pane.

It is important to know that data best connects to tableau from spreadsheets if it 'has not been severely formatted' in excel by adding borders, etc., but left in its raw form. The data from the required fields or columns or rows needed for analysis can be selected by dragging and dropping them in the analysis fields, with the result displayed without requiring any calculations or plots because they are done spontaneously as the needed data is selected in a point and click or drag and drop fashion [42].

Many organisations use tableau to improve their stance in every business situation [15], accessories like dynamic dashboards make it possible to monitor depreciating product sales or services much more easily and it takes data summarisation to another level from what can be derived from pivot tables, used for flexible reporting. To reiterate the significance of visualisation in appreciating data, the works of [43, 44] opines that visual aids in analytics makes it possible to access new and in-depth understandings from data, by helping to transform unprocessed data into knowledge that can be leveraged on to support organisations in policy- and decision-making. [44] further stated that visualisation by itself does not really do justice to data, compared to when used together with other analytic techniques—benefits SAS and other tools can offer such as Monte Carlo simulation, data modelling, data analysis, etc.

7   The BI Tool Qlikview

Qlikview is a product of the BI company Qliktech, and it is also a data visualisation tool like tableau said to be a 'commercial BI platform' by [45] though [46] argued that Qlikview is simply a business discovery tool and not a BI tool. However, ultimately, Qlikview helps turn data into knowledge and that in itself is BI.

Qlikview is distinct to its likes in that, it has an in-memory data storage and it can refresh data in real time in an operational BI active environment to support financial transactions. It is also easy to use for tech experts and non-tech experts alike. Qlikview produces highly dynamic dashboard, applications and reports by design, and can combine data from different sources such as Excel, Oracle, SQL server, etc. It can also work as a mini search engine known as associative search, for the data within it and spotting relationships or associations between data as it goes [45].

The dashboard is also a major function of Qlikview, that does the back end analytics, and at the same time serves as the user interface as it displays all information relating to the data inside it. Such that inefficiencies can be spotted easily [47], saving time, and avoiding situations where irreparable damages have been done before it can be noticed.

For organisations, Qlikview can consolidate business information across different regions and make organisation-wide performance monitoring possible on a single screen per time without having to wait for end of day or end of month reporting.

8 SAS Enterprise Guide

According to [41], SAS Enterprise Guide 'is a point-and-click, menu-and-wizard-driven tool that empowers users to analyse data and publish results'.

This software and product of the SAS institute has the capability to access, query and manage wide range of data sets, which makes for flexible and extensive data manipulation by the SAS user. SAS Enterprise Guide is compatible with the widely used Microsoft windows for analysis and improves the usability of SAS among many users by its development. SAS Enterprise Guide is widely used for advanced level of analytics, business intelligence, predictive analytics, general data management and analytics involving multiple variables.

SAS enterprise guide is usable on different systems such as windows server, personal computers and UNIX server, in many instances and has simple to learn packages.

SAS enterprise guide receives data for reporting into itself in any format especially excel which is the most common 'mathematical' tool known to most, who may be unaware of advanced analytic tools.

Reports on SAS can be prepared and viewed in HTML—HyperText Markup Language format—for when results are going directly on a website, in PDF—Portable Document Format: for when reports are sent off to stakeholders directly. On SAS enterprise guide, RTF—Rich text formats, text, SAS report format are other ways to view reports. Reporting and analysis can start and end on SAS enterprise guide as reports or sent off directly for management information without the use of another software; it is also a self-service tool [40, 41].

9 Analytics in Banking

Banks have always been the backbone of economies in both developed and developing countries, and in order to balance the economy ensure banks continuity there has to be conscious data optimisation efforts directed to banks.

According to [48] when analytics is employed in the banking industry, the true state of things are known and positives are not confused for negatives and vice versa, which is the crux of Bankhawk analytics work in Dublin [49]. Bankhawk analytics, whose goal is to assist financial organisations become more profitable and competitive, granted an interview to gtnews, the founder and chief executive officer of Bankhawk analytics pointed out that banks are now gearing up to the fact that they have a lot to gain from the use of technology and are now coming to the realisation of the vast amounts of data they possess that remains untapped. He also mentioned that; to fully understand organisations; who are customers to banks; the secret is in understanding; their income, their expenses and their cashflow. However, this is to be done in the context of business analytics and the use of the necessary tools, an opinion that [50] shares as well, stating:

Data and Analytics is allowing financial services firms to take a far more holistic view of how their businesses are performing, and providing more complete and insightful to support strategic decision making.

As an analytics tool, Tableau helps increase speed and data understanding and has helped global financial service providers like Wells fargo, Barclays, Citibank, Societe generale better prepare their data for analysis and get insights from the data in an efficient manner than before the use of this tool was employed, by reducing the time involved in reporting considerably—turning what could take weeks into days, days into minutes, etc.—it has also served Ernst and Young in managing risks and its effect within organisations is getting everyone on the same page [15]. The benefits of a tool like this cannot be over emphasied for banks with an enormous responsibility to economies, another advantage of tableau is that it can be used offline on already extracted data.

Having analytics in banking as [22] said would allow the banks information to be accessed from a single source by those employees who need it, and as noted by [6] banks are quick to adopt IT innovations for their processes and analytics techniques presently used in accessing loan approval and detecting fraud, this makes it possible for analytics to be extended to other risk management functions such as loan review, by making it possible for the banks to rely on fact-based statistics that they can work with, there by reducing risks.

Quoting the Financial stability board and IMF report, [44] noted that good analysis and good data is paramount in financial surveillance both locally and internationally.

The studies of [5] about social media text analytics, which was tried on JP morgan chase bank, lets the banks know about their environment in terms of competition and risk analysis including internal and external.

Notwithstanding the steps that a few banks are taking with the use of BI, there is yet so much more to be exploited in analytics within banking especially in areas where the data exists, only waiting to be exploited so as to make each activity/unit profitable or productive with the quality of outputs they give the organisation. For banks one of such is what [51] calls Lending management intelligence, a service they render to banks which evaluates loan timings, repayment schedules and also maintains updated reports on existing loans being reviewed, and bad or non-performing loans.

10   Conclusions and Discussion

Having defined what big data is and some tools that can aid organisations in benefitting from big data, it is important to mention some challenges to the exploration of big data.

Some of the major challenges to understanding big data is catching up to the rate at which data is produced. Also, organisations are yet to fully grasp the capture, storages and analysis of unstructured data—which is what big data mainly contains, and its uses and implementation [4].

To overcome these challenges, [3] highlighted four strategies as follows:

Performance management: which involves analysis of business' transactional data, which is, consolidating all businesses internal data and creating a holistic view with which insights can be generated.

Data exploration: this entails the use of statistics on existing data to predict future outcomes by studying past consumer behaviour to reveal answers that have not been thought of, in order to retain customers, prevent attrition, engage in target marketing and cross and upsell.

Social analytics: involves using non-transactional data to gather beneficial insights for organisations from social media sites/apps, measuring engagements/views on videos as well as conversion rates.

Decision science: this also involves non-transactional data on social media, but not for analysis but to identify and decide the best course of action from what consumers resonates with, judging from their engagement with certain posts or campaigns.

Now, using all the tools that can provide organisations the benefits listed above involves some level of expertise which many organisations may not possess, which is why [2] identified three things that can be done to fully engage big data; organisations need to Educate willing members of the organisation who have the potential to carry out these activities, and if such people are not within an organisations, they are to be Acquired, and finally these set of individuals are to be Empowered in orther to be able to perform their duties, within organisational reason.

Although big data analytics is a new terrain of analysing data that seems to be 'out of control', the benefits of gaining insight into what exactly can improve profit for the organisation, and reduce errors—being intentional and exact with business strategies and policies, knowing what individual customers are willing to buy and continue buying is really priceless, which is what ecommerce sites like Amazon, Ebay and the likes are using to become Internet business giants.

Following the tools that were mentioned as useful in big data harnessing in earlier sections of this work such as Hadoop, etc., it is however highly recommended that businesses, including banks and financial institutions plan the analytical process to include the right people with the right skills sufficient to handle tools for big data manipulation. It is also important to access the value big data or BI tools can afford the business as time goes on and as the business grows. More importantly businesses are encouraged to start small with analytics and 'simple big data'.

References

1. Patanakul, P., & Shenhar, A. J. (2012). What project strategy really is: The fundamental building block in strategic project management. *Project Management Journal, 43*(1), 4–20.
2. Elias, H. (2012). The big data challenge: how to develop a winning strategy. http://www.cio. com/article/2395010/data-management/the-big-data-challenge–how-to-develop-a-winning-strategy.html.
3. Parise, S., Iyer, B., & Vesset, D. (2012). Four strategies to capture and create value from big data. Ivey Business Journal.
4. Oracle. (2015). Integrate for insight: Combining big data tools with traditional data management offers. http://www.oracle.com/us/technologies/big-data/big-data-strategy-guide-1536569.pdf.
5. Ribarsky, W., Wang, D. X., & Dou, W. (2014). Social media analytics for competitive advantage. *Computers & Graphics, 38,* 328–331.
6. Moro, S., Cortez, P., & Rita, P. (2015). Business intelligence in banking: A literature analysis from 2002 to 2013 using text mining and latent Dirichlet allocation. *Expert Systems with Applications, 42*(3), 1314–1324.
7. Haque, W., & Dhanoa, R. (2015). A framework for reporting and analytics. s.l., Academic Search Complete.
8. Dull, T. (2016). *A non-geek's big data playbook: Hadoop and the enterprise data warehouse.* Cary: NC USA, SAS Institute Inc.
9. Pearson Education Inc. (2014). *Foundations of business intelligence: Databases and information management.* s.l.: Pearson Education Inc.
10. Bacic, D., & Fadlalla, A. (2016). Business information visualisation intellectual contributions: An integrative framework of visualisation capabilities and dimensions of visual intelligence. *Decision Support Systems, 89,* 77–86.
11. Ghazanfari, M., Jafari, M., & Rouhani, S. (2011). A tool to evaluate the business intelligence of enterprise systems. *Scientia Iranica, 18*(6), 1579–1590.
12. Xia, B. S., & Gong, P. (2012). Review of business intelligence through data analysis. *Benchmarking: An international Journal, 21*(2), 300–309. (Ramachandran, 2016).
13. Nadeem, M., & Jaffri, S. A. H. (2004). *Application of business intelligence in banks (Pakistan).* Karachi, Pakistan: Cornell University Library.
14. SAP. (2016). Increase business agility with the right information, when and where it's needed. Retrieved July 26, 2016, http://go.sap.com/uk/documents/2016/06/e65abd4e-777c-0010-82c7-eda71af511fa.html.
15. Tableau. (2016). Banking. Retrieved July 28, 2016, www.tableau.com/stories/topic/banking.
16. Olap.com. (2016). Business performance management. Retrieved August 4, 2016, http://olap. com/learn-bi-olap.
17. Omotayo, F. O. (2015). Knowledge management as an important tool in organisational management: A review of literature. *Library Philosophy and Practice* 1–23. General OneFile, EBSCOhost.
18. Larson, D., & Chang, V. (2016). A review and future direction of agile, business intelligence, analytics and data science. *International Journal of Information Management, 36*(5), 700–710.
19. Florin, D., & Radu-Daniel, L. (2014). The impact of business intelligence recognition of goodwill valuation. *Procedia Economics and Finance, 15,* 1779–1786.
20. Chen, H., Chiang, R. H. L., & Storey, V. C. (2012). Business intelligence and analytics: From big data to big impact. *MIS Quarterly 36*(4), 1165–1188.
21. Turban, E., Sharda, R., & Delen, D. (2011). *Decision support and business intelligence systems* (9th ed.). Upper Saddle River, NJ: Prentice-Hall.
22. Rao, G. K., & Kumar, R. (2011). Framework to integrate business intelligence and knowledge management in banking industry. *Review of Business and Technology Research, 4*(1).

23. Beal, V. (2015). BPM—Business performance management. Retrieved August 4, 2016 http://www.webopedia.com/TERM/B/BPM.html.

24. Ramachandran, M. (2016). Introduction to BPMN for BI and BD Analytics. In *A lecture on Business Process re-engineering*. Leeds Bekett University. February.

25. Wong, W. P., Ahmad, N. H., Nasurdin, A. M., & Mohamad, M. N. (2014). The impact of external environmental on business process management and organizational performance. *Service Business, 8*(4), 559–586.

26. Web Finance Inc. (2016). Business performance management (BPM). Retrieved August 4, 2016, http://www.businessdictionary.com/definition/business-performance-management-BPM.html.

27. Adrian. (2014). User interface design examples for your inspiration. Retrieved August 3, 2016, http://designmodo.com/user-interface-design-examples/.

28. Cartouche. (2012). Types of user interfaces. Retrieved August 3, 2016, http://www.e-cartouche.ch/content_reg/cartouche/ui_access/en/html/UnitGUI_UI.html.

29. Xerox. (2016). 35 Interface Innovations that Rocked Our World. Retrieved August 3, 2016, https://www.xerox.com/en-us/insights/user-interface-examples.

30. Vartak, M., et al. (2015). SEEDB: Efficient data-driven visualisation recommendations to support visual analytics. *Proceeedings of the VLDB Endowment, 8*(13), 2182–2193.

31. Beal, V. (2016). Business analytics. Retrieved August 4, 2016, http://www.webopedia.com/TERM/B/business_analytics.html.

32. Rouse, M. (2010). Business analytics (BA). Retrieved August 4, 2016, http://searchbusinessanalytics.techtarget.com/definition/business-analytics-BA.

33. SAP. (2014). BI Strategy. Retrieved August 4, 2016, https://www.sapbi.com/wp-content/themes/sapbi/library/images/bistrategy/BI%20Strategy.pdf.

34. Sherman, R. (2015). Understanding BI analytics tools and their benefits. Retrieved August 8, 2016, http://searchbusinessanalytics.techtarget.com/feature/Understanding-BI-analytics-tools-and-their-benefits.

35. Software Advice. (2016). Compare UK Business Intelligence (BI) Software. Retrieved August 8, 2016, http://www.softwareadvice.com/uk/bi/.

36. Alade, A. T. (2016). *Research Proposal Assessment 3*. s.n.: Leeds.

37. Baiju, N. (2014). Top business intelligence (BI) tools in the market. Retrieved August 8, 2016, http://bigdata-madesimple.com/top-business-intelligence-bi-tools-in-the-market/.

38. DISYS. (2016). Banks, financial institutions can leverage buisiness intelligence to making lending decisions. Retrieved July 28, 2016, www.disys.com/banks-financial-institutions-can-leverage-business-intelligence-to-make-lending-decisions/.

39. DISYS. (2016). Business intelligence can help banks improve mobile solutions. Retrieved July 28, 2016, http://www.disys.com/business-intelligence-can-help-banks-improve-mobile-solutions/.

40. McDaniel, S., & Hemedinger, C. (2007). SAS for Dummies. Indiana: Wiley Publishing, Inc.

41. SAS Institute Inc. (2016). SAS the power to know. Retrieved September 5, 2016, http://www.sas.com/en_gb/home.html.

42. Tech Analytics. (2015). 1 Basics connecting to data—Tableau for beginners. s.l.: YouTube.

43. Boaler, J. (2016). Seeing as understanding: The importance of visual mathematics. Retrieved August 4, 2016, https://bhi61nm2cr3mkdgk1dtaov18-wpengine.netdna-ssl.com/wp-content/uploads/2016/04/Visual-Math-Paper-vF.pdf.

44. Flood, M. D., Lemieux, V. L., Varga, M., & Wong, B. W. (2015). The application of visual analytics to financial. *Journal of Financial Stability*. ScienceDirect, EBSCOhost.

45. Mohamed, A. (2011). Review: QlikTech's QlikView business intelligence platform. Retrieved August 8, 2016, http://www.computerweekly.com/feature/Review-QlikTechs-QlikView-business-intelligence-platform.

46. Data & Tools. (2013). QlikView Introduction. s.l.:s.n.

47. Visual Intelligence, 2016. What is QlikView? Retrieved August 8, 2016, https://www.visualintelligence.co.nz/qlikview/.

48. IBM Analytics. (2016). Finance in focus: Turbocharge your anti–money laundering program. Retrieved July 26, 2016, http://www.ibmbigdatahub.com/podcast/finance-focus-turbocharge-your-anti-money-laundering-program.
49. Bankhawk Analytics. (2016). Banking analytics: The insights corporates need. Retrieved July 27, 2016, www.bankhawk.com/updates/banking-analytics-the-insights-corporates-need-2/.
50. PWC. (2015). Data and analytics in the banking and capital markets sector. Retrieved August 3, 2016, http://www.pwc.co.uk/data-analytics/industries/banking-and-capital-markets.html.
51. Corporate Technologies, 2016. Banking. Retrieved July 28, 2016, www.cptech.com/banking-it-business-intelligence/.

Index

© Springer International Publishing AG 2017
A. Hosseinian-Far et al. (eds.), *Strategic Engineering for Cloud Computing and Big Data Analytics*, DOI 10.1007/978-3-319-52491-7

Printed in the United States
By Bookmasters